计算机科学与技术专业实践系列教材

基于工作任务的
Java程序设计实验教程

宋锋 冯君 崔蕾 主编
谭业武 刘春霞 刘启明 李宏光 副主编

清华大学出版社
北京

内 容 简 介

本书由两篇内容组成,第一篇内容是与理论教材配套的实验内容,特点是采用任务驱动的方式来进行组织,每一章都包括几个与 Java 理论知识以及生活密切相关的实例的练习,每一个任务都有详细的实施步骤,读者通过循序渐进的练习,达到掌握 Java 语言的知识点、积累开发经验的目的。

第二篇内容是一个"图书管理系统"的综合应用案例,综合应用了 Java 的基本语法、Java 的程序控制结构、面向对象的分析设计、数据库、图形用户界面、异常处理、JDBC 等的相关知识,可以用作为 Java 程序设计配套的课程设计参考,也可以作为 Java 团队合作项目的参考。

本书可作为高等学校计算机及相关专业的 Java 程序设计课程的实验教材,也可供从事 Java 程序开发的技术人员参考。

本书封面贴有清华大学出版社防伪标签,无标签者不得销售。
版权所有,侵权必究。举报: 010-62782989, beiqinquan@tup.tsinghua.edu.cn。

图书在版编目(CIP)数据

基于工作任务的 Java 程序设计实验教程/宋锋等主编. —北京: 清华大学出版社,2015(2024.9重印)
计算机科学与技术专业实践系列教材
ISBN 978-7-302-40898-7

Ⅰ. ①基… Ⅱ. ①宋… Ⅲ. ①JAVA 语言-程序设计-高等学校-教材 Ⅳ. ①TP312

中国版本图书馆 CIP 数据核字(2015)第 165482 号

责任编辑: 白立军
封面设计: 傅瑞学
责任校对: 李建庄
责任印制: 丛怀宇

出版发行: 清华大学出版社
    网 址: https://www.tup.com.cn, https://www.wqxuetang.com
    地 址: 北京清华大学学研大厦 A 座   邮 编: 100084
    社 总 机: 010-83470000   邮 购: 010-62786544
    投稿与读者服务: 010-62776969, c-service@tup.tsinghua.edu.cn
    质量反馈: 010-62772015, zhiliang@tup.tsinghua.edu.cn
    课件下载: https://www.tup.com.cn,010-62795954
印 装 者: 三河市人民印务有限公司
经 销: 全国新华书店
开 本: 185mm×260mm   印 张: 19   字 数: 473 千字
版 次: 2015 年 8 月第 1 版   印 次: 2024 年 9 月第 8 次印刷
定 价: 49.00 元

产品编号: 062489-02

序

当今世界,以计算机技术、通信技术和控制技术为代表的 3C 技术正迅猛发展,而以 Internet 为代表的全球范围内信息基础设施的建设成就,标志着人类社会已进入信息时代。应用型人才培养是社会发展和高等教育发展的必然要求,经济社会发展迫切需要高等学校培养出在知识、能力、素质等诸方面都适应社会需要的不同层次的应用型人才,满足信息化社会建设的需求。实施高等教育名校建设工程,对大力发展高等教育,指导高等教育特色发展,全面提高教育质量,增强高等教育的竞争力和服务经济社会能力具有重大意义。

罗伯特·加涅是美国教育心理学家,加涅将认知学习理论应用于教学过程,加涅理论中的引出作业和提供反馈是一种教学策略。从"应用型"人才培养的角度来说,学生的实践能力提升是一个重要问题,需要学校和教师采取一些有效手段来增强学生的实践能力,树立以学生为本的观念,尊重学生的个性特点,因材施教,增加学生对于课程、专业的选择空间。

本套教材(指刘启明教授参与编写的教材)是我们多年来进行"应用型人才培养教学内容、课程体系改革"的综合成果。我们提出的课程内容设置方案,目的是推进人文与自然的融合,适应学生能力、兴趣、个性、人格全面发展的需要,强化学生的实践能力和创新能力培养。实施教学方式、教学内容、考核机制的全面改革,在培养学生的信息能力和信息素养方面具有先导作用,为计算机课程教学内容、课程体系改革,设计了一个全新的框架。

本套教材以应用型、技能型人才培养为目标,以重点专业建设为平台,围绕着教育教学改革、创新人才培养、提高人才培养质量的教育发展理念展开。每部教材都是由应用型名校计算机专业课教师或者计算机实验教学示范中心专业教师编写完成的。

在本套教材的编写过程中,我们得到许多专家的精心指点和热情帮助。教育部计算机科学与技术教学指导委员会先后三次在我校召开计算机基础课程教学研讨会,清华大学、北京大学、中国人民大学、复旦大学、浙江大学、南京大学、中国科学技术大学等近百所高校的老师参加。专家学者对本套教材的编写提出了很多宝贵意见。

本套教材的出版得到清华大学出版社的大力支持,正是他们精益求精的工作,才使这一系统工程得以顺利完成,并得到高度评价,在此表示衷心感谢。

刘启明
2015 年 6 月

前　　言

　　Java 是为了适应智能设备和网络应用而产生的一种程序设计语言,拥有面向对象、跨平台、高性能、分布性和可移植性等特点,是目前被广泛使用的编程语言之一,近些年来的编程语言排名中,Java 语言一直位列第一位。Java 可以用于传统的桌面应用程序的编程,也可以用于家电、智能设备、手持设备、通信设备等嵌入式应用程序的开发,随着网络向着云计算、物联网的方向发展,Java 语言具有更加广阔的应用市场和应用前景。社会对 Java 工程师的需求量一直很大,掌握 Java 语言,能够进行典型的 Java 应用程序的开发,是对普通高等院校计算机及相关专业学生最基本的能力要求之一。

　　本书通过通俗易懂的语言和实用生动的例子,以任务驱动的方式带领读者进行上机实验,每个任务都有详细的实施步骤,方便老师和同学操作检验,任务还备有思考、讨论或是任务扩展,使读者能在掌握基本知识点的基础上,达到能举一反三的目的。

　　全书共分为两篇,第一篇为基本实验篇,第二篇为综合应用实例。

　　第一篇由 16 章组成。第 1 章通过 4 个任务,介绍 JDK 环境的安装、配置,使用记事本进行 Java 程序的开发,在控制台中使用 JDK 环境对 Java 程序进行编译和运行。第 2 章通过 5 个任务,介绍常量和变量的定义、取值范围、表达式、转义字符、数据类型转换、注释的使用方法。第 3 章通过 6 个任务,介绍了 if…else 及 switch…case 选择分支结构的使用方法。第 4 章通过 6 个任务的练习,介绍了 for 循环、while 循环、do…while 循环的语法和使用方法,以及结束循环的方法。第 5 章通过 3 个任务,对 Java 中的数组的定义和使用方法进行了详细的阐述。第 6 章通过 3 个任务,介绍类和对象的概念、定义和使用方法。第 7 章通过 3 个任务,介绍继承的概念和在程序中的使用方法。第 8 章通过 2 个任务,介绍多态的特点和使用方法。第 9 章通过 2 个任务,阐述接口的特点和使用方法。第 10 章通过 5 个任务,介绍异常的定义、异常的处理方法、自定义异常及使用方法。第 11 章通过 5 个任务,介绍使用图形用户界面开发桌面应用的方法。第 12 章通过 4 个任务,介绍输入输出流的使用方法。第 13 章通过 2 个任务,介绍了 List 集合和 Map 集合的使用方法。第 14 章通过 5 个任务,介绍网络编程中常用对象的使用方法。第 15 章通过 4 个任务,介绍多线程的特点和使用方法。第 16 章通过"会员管理信息系统"的开发,介绍纯 JDBC 驱动连接与操作数据库中数据的方法。

　　第二篇内容是一个"图书管理系统"的综合应用案例,综合应用了 Java 的基本语法、Java 的程序控制结构、面向对象的分析设计、数据库、图形用户界面、异常处理、JDBC 等的相关知识,可以用作 Java 程序设计配套的课程设计参考,也可以作为 Java 团队合作项目的参考。

　　由于作者水平有限,缺点和欠妥之处难免,恳请读者帮助指正。

作　者

2015 年 6 月

目　　录

第一篇　基础实验篇

第1章　打开Java之门 ··· 3
  1.1　实验目的 ··· 3
  1.2　实验任务 ··· 3
  1.3　实验内容 ··· 3
    1.3.1　任务1　使用记事本编写Java程序并编译运行 ············· 3
    1.3.2　任务2　联合编译运行多个Java程序 ······················ 6
    1.3.3　任务3　使用Eclipse集成开发平台开发简单Java程序 ······· 8
    1.3.4　任务4　使用Eclipse集成开发平台联合运行Java程序 ······ 12

第2章　Java的基本语法 ··· 14
  2.1　实验目的 ·· 14
  2.2　实验任务 ·· 14
  2.3　实验内容 ·· 14
    2.3.1　任务1　编写程序显示各个数值数据类型的最值 ············ 14
    2.3.2　任务2　声明不同类型的变量并进行赋值输出 ·············· 15
    2.3.3　任务3　根据输入圆半径的值求圆的面积 ·················· 17
    2.3.4　任务4　从键盘输入3个数并求它们的平均数 ··············· 18
    2.3.5　任务5　编写程序查看常用转义字符的效果 ················ 19

第3章　选择结构 ··· 21
  3.1　实验目的 ·· 21
  3.2　实验任务 ·· 21
  3.3　实验内容 ·· 21
    3.3.1　任务1　判断键盘输入的数据是否能被7整除 ··············· 21
    3.3.2　任务2　成绩等级判断 ·································· 23
    3.3.3　任务3　判断键盘输入的数据是否为大写字母 ·············· 25
    3.3.4　任务4　计算销售提成 ·································· 26
    3.3.5　任务5　判断回文数 ···································· 28
    3.3.6　任务6　根据订单状态标志显示订单状态 ·················· 31

第4章　循环结构 ··· 34
  4.1　实验目的 ·· 34
  4.2　实验任务 ·· 34
  4.3　实验内容 ·· 34

| 4.3.1 任务1 摄氏温度到华氏温度的转换表 ·· 34
| 4.3.2 任务2 抽奖 ·· 35
| 4.3.3 任务3 求和 ·· 37
| 4.3.4 任务4 break 和 continue ·· 37
| 4.3.5 任务5 猜数字游戏 ··· 38
| 4.3.6 任务6 马克思手稿中的数学题 ··· 40

第5章 数组 ·· 42
  5.1 实验目的 ·· 42
  5.2 实验任务 ·· 42
  5.3 实验内容 ·· 42
        5.3.1 任务1 成绩统计 ·· 42
        5.3.2 任务2 食堂饭菜质量评价 ··· 44
        5.3.3 任务3 打印杨辉三角形 ··· 45

第6章 类和对象 ·· 47
  6.1 实验目的 ·· 47
  6.2 实验任务 ·· 47
  6.3 实验内容 ·· 47
        6.3.1 任务1 手机类的封装 ·· 47
        6.3.2 任务2 基于控制台的购书系统 ··· 50
        6.3.3 任务3 简单投票程序 ·· 55

第7章 继承 ·· 58
  7.1 实验目的 ·· 58
  7.2 实验任务 ·· 58
  7.3 实验内容 ·· 58
        7.3.1 任务1 公司雇员类封装 ··· 58
        7.3.2 任务2 汽车租赁系统 ·· 62
        7.3.3 任务3 饲养员喂养动物 ··· 67

第8章 多态 ·· 71
  8.1 实验目的 ·· 71
  8.2 实验任务 ·· 71
  8.3 实验内容 ·· 71
        8.3.1 任务1 图形面积周长计算小程序 ·· 71
        8.3.2 任务2 饲养员喂养动物程序优化 ·· 74

第9章 接口 ·· 78
  9.1 实验目的 ·· 78
  9.2 实验任务 ·· 78
  9.3 实验内容 ·· 78

	9.3.1	任务 1	设计实现发声接口	78
	9.3.2	任务 2	动物乐园	81

第 10 章 异常处理 … 86
10.1 实验目的 … 86
10.2 实验任务 … 86
10.3 实验内容 … 86
    10.3.1 任务 1 判断从键盘输入的整数是否合法 … 86
    10.3.2 任务 2 处理除数为 0 的异常 … 87
    10.3.3 任务 3 处理数组的下标越界异常 … 88
    10.3.4 任务 4 特殊字符检查器 … 89
    10.3.5 任务 5 使用 try-with-resource 进行读取文件处理 … 91

第 11 章 图形用户界面设计 … 93
11.1 实验目的 … 93
11.2 实验任务 … 93
11.3 实验内容 … 93
    11.3.1 任务 1 公司员工信息录入程序 … 93
    11.3.2 任务 2 小学生习题训练程序 … 96
    11.3.3 任务 3 "我所喜爱的主食和副食"问卷调查 … 101
    11.3.4 任务 4 员工信息处理菜单 … 104
    11.3.5 任务 5 商场收银软件 … 106
11.4 课后巩固练习 … 111

第 12 章 输入输出流 … 112
12.1 实验目的 … 112
12.2 实验任务 … 112
12.3 实验内容 … 112
    12.3.1 任务 1 FileWriter 和 BufferedWriter 比较 … 112
    12.3.2 任务 2 给源程序加入行号 … 114
    13.3.3 任务 3 统计英语短文字母 A 出现的次数 … 115
    13.3.4 任务 4 简易 Java 考试系统 … 117

第 13 章 Java 集合框架 … 128
13.1 实验目的 … 128
13.2 实验任务 … 128
13.3 实验内容 … 128
    13.3.1 任务 1 使用 List 模拟图书系统实现歌曲的增、删、改、查 … 128
    13.3.2 任务 2 使用 Map 模拟电话号码管理程序 … 132

第 14 章 Java 网络编程 … 137
14.1 实验目的 … 137

14.2	实验任务		137
14.3	实验内容		137
	14.3.1	任务1 显示URL对象的相关属性	137
	14.3.2	任务2 获取本机和远程服务器地址的方法	139
	14.3.3	任务3 检查本机指定范围内的端口是否已经使用	140
	14.3.4	任务4 使用TCP通信编写聊天软件	141
	14.3.5	任务5 使用UDP通信编写聊天程序	144

第15章 多线程 ... 149

15.1	实验目的		149
15.2	实验任务		149
15.3	实验内容		149
	15.3.1	任务1 使用Thread和Runnable模拟时钟线程	149
	15.3.2	任务2 线程控制的基本方法	151
	15.3.3	任务3 模拟夫妻二人去银行取钱	155
	15.3.4	任务4 生产者-消费者问题	157

第16章 数据库操作 ... 163

16.1	实验目的	163
16.2	实验任务	163
16.3	实验内容	163

第二篇 综合实例篇

第17章 图书管理系统 ... 179

17.1	图书管理系统业务需求分析		179
	17.1.1	系统使用对象分析	179
	17.1.2	业务需求分析	179
	17.1.3	系统功能模块分析	180
	17.1.4	系统数据库分析	180
17.2	功能模块实现		181
	17.2.1	用户登录模块设计	181
	17.2.2	用户管理模块设计	187
	17.2.3	用户密码管理模块设计	197
	17.2.4	读者信息管理模块设计	199
	17.2.5	图书信息管理模块设计	226
	17.2.6	图书借阅/归还操作模块设计	232
	17.2.7	罚款管理模块设计	260
	17.2.8	报表打印模块设计	267

17.2.9　帮助管理模块设计 ……………………………………………………… 273
　　17.2.10　主界面管理模块设计 …………………………………………………… 276
17.3　系统发布与总结 ………………………………………………………………… 289
　　17.3.1　项目打包 …………………………………………………………………… 289
　　17.3.2　项目总结 …………………………………………………………………… 291

第一篇 基础实验篇

本篇内容与教材对应,通过各个章节、各个任务循序渐进的练习,使读者能够掌握 Java 程序的编写、编译的开发方法,Java 集成开发环境 Eclipse 的使用方法,Java 的基础语法、标识符的定义方法,Java 的程序控制结构,数组的定义与使用方法,类和对象的定义、接口的定义,类的封装、继承、多态的定义、特点及应用方法,Java 中的异常处理方法,图形用户界面的组件、布局、界面元素及界面的设计方法,输入输出流的应用,Java 集合框架的定义及应用,Java 网络编程的知识,Java 多线程,使用 JDBC 驱动连接数据库进行数据操作的知识等。

每章都由实验目的、实验任务、实验内容 3 个部分组成,实验目的是练习完成本章的实验后所能掌握的知识点;实验任务是与理论知识以及实际生活相关的案例任务,每个任务都描述了完成这个任务所能掌握的知识点、该任务的实施步骤、运行结果,大部分任务还有思考、讨论和任务拓展,通过本篇各个章节的练习,使读者能在掌握基本知识点的基础上,达到能够举一反三的目的。

第1章 打开 Java 之门

1.1 实验目的

(1) 掌握 JDK 的安装与配置。
(2) 掌握使用记事本编写 Java 程序的方法。
(3) 掌握使用控制台编译运行 Java 程序的方法。
(4) 掌握 Eclipse 的下载及安装方法。
(5) 掌握使用 Eclipse 集成开发平台编写运行 Java 程序的方法。

1.2 实验任务

(1) 任务1：使用记事本编写 Java 程序并编译运行。
(2) 任务2：联合编译运行多个 Java 程序。
(3) 任务3：使用 Eclipse 集成开发平台开发简单 Java 程序。
(4) 任务4：使用 Eclipse 集成开发平台联合编译 Java 程序。

1.3 实验内容

1.3.1 任务1 使用记事本编写 Java 程序并编译运行

1. 任务目的

(1) 掌握使用记事本编写 Java 程序的方法。
(2) 掌握在控制台中使用 JDK 编译与运行 Java 程序的方法。

2. 任务描述

在 JDK 开发环境下载、安装与配置(关于 JDK 的下载、安装与配置请参考课本)完成后,使用记事本编写一个程序 HelloJava.java,在控制台运行程序后,显示"Hello Java World,I'm coming!"。

3. 实施步骤

1) 新建文件

在指定的文件夹(如本例所用 D:/Lab1/task1,在完成本任务时,读者可以把文件放在任何你能记清楚路径的文件夹中)中新建一个文本文档,创建方法为：在资源管理器中右击,在弹出的快捷菜单中选择"新建"→"文本文档"选项,如图 1-1 所示。

修改文本文档的名字为 HelloJava.java,打开这个文本文档后输入如图 1-2 所示的内容。

HelloJava.java 的代码如下：

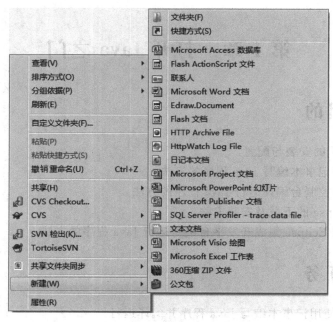

图 1-1 新建文本文档

图 1-2 HelloJava.java 文件的内容

```
public class HelloJava{
    public static void main(String[] args){
        System.out.println("Hello Java World,I'm coming!");
    }
}
```

2）打开控制台 cmd 程序并切换路径到 HelloJava.java 所在的路径

单击系统桌面左下角的"开始"按钮，选择"所有程序"→"附件"→"命令提示符"选项，或执行"开始"→"运行"命令，打开"运行"窗口，在"运行"窗口框中，使用 cmd 命令（Windows XP 系统、Windows 2003 系统），或在"开始"→"查询"框中使用 cmd 命令，将会打开 DOS 命令环境（也称为控制台环境），在 DOS 命令行环境中通过 DOS 命令将路径切换到 HelloJava.java 文件所在的路径，如图 1-3 所示。

从图 1-3 中可以看出，切换路径时，要在控制台窗口中输入数据，当路径比较长，或者路径中含有中文时，从键盘上一个字符接着一个字符地输入，会比较麻烦，有没有比较简单一点的方法呢？下面介绍一种快捷的方法。

首先在资源管理器中找到 HelloJava.java 所在的路径，将地址栏中的路径复制下来，然

图 1-3 在 DOS 控制台将路径切换到当 HelloJava.java 文件所在的路径

后在 DOS 控制台窗口标题栏上右击,在弹出的快捷菜单中选择"编辑"→"粘贴"(或者直接在 DOS 窗口中右击)选项,则会将路径信息瞬间输入到 DOS 控制台中去。在标题栏上右击进行粘贴方法如图 1-4 所示。

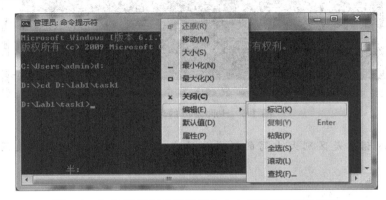

图 1-4 在 DOS 控制台标题栏上右击后进行粘贴的方法

3) 使用 JDK 编译器 javac 进行编译

在控制台中的命令行提示符 D:\Lab1\task1> 的后面输入如下代码:

```
javac HelloJava.java
```

输入完成后,按 Enter 键,若出现如图 1-5 编译完成的效果(空一行后,还是出现命令行提示符 D:\Lab1\task1>),则编译成功。

图 1-5 HelloJava.java 编译成功后的 DOS 控制台界面

此时到 HelloJava.java 所在的文件夹(本任务位置为 D:\Lab1\task1)中去查看,会看到 HelloJava.class,这就是 Java 程序在经过 javac 编译工具编译后生成的字节码文件。

4）运行 Java 程序

在图 1-5 所示的控制台中继续输入如下代码：

```
java HelloJava
```

输入完成后，按 Enter 键，则会运行上面编译生成的 HelloJava.class 字节码文件，出现如图 1-6 所示的程序运行效果。

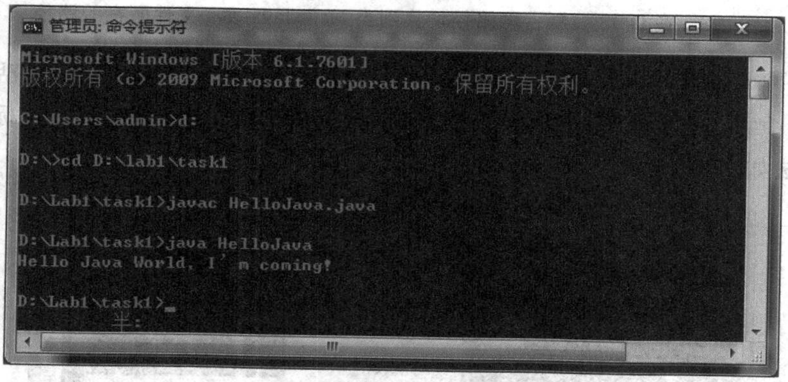

图 1-6 HelloJava 的运行效果

注意：Java 语言是区别大小写的强类型语言，因此在控制台环境中使用如下两个命令时：

```
javac HelloJava.java
java HelloJava
```

注意 javac 和 java 的参数：HelloJava 中的字符一定要严格区分大小写，否则 Java 程序会把它当成一个新的变量，如 HelloJava、helloJava、Hellojava 将会被当成 3 个变量对待。

4. 任务拓展

上面的例子实现了向 Java 世界问好的功能，如何实现向自己问好的功能呢？

1.3.2 任务2 联合编译运行多个 Java 程序

1. 任务目的

掌握在控制台中联合编译并运行多个 Java 程序的方法。

2. 任务描述

在实际应用中，一个应用程序可能会由多个文件组成，这样的应用程序中包括入口程序（带有 main()方法）、功能程序（实现各个功能）。

本任务的主程序为 MyHello.java，在程序运行后，会依次调用功能程序 Hello1.java（在方法中显示"功能程序1"）、Hello2.java（在方法中显示"功能程序2"）、Hello3.java（在方法中显示"功能程序3"）。

3. 实施步骤

1）创建主程序

按任务1中创建与编写 Java 程序的方法，创建文件夹 D:/Lab1/task2，创建主程序

MyHello.java,编写程序代码如下:

```java
public class MyHello{
    public static void main(String[] args){
        Hello1 hello1=new Hello1();        //第 1 个功能类实例
        hello1.show();                      //第 1 个功能类中的方法
        Hello2 hello2=new Hello2();        //第 2 个功能类实例
        hello2.show();                      //第 2 个功能类中的方法
        Hello3 hello3=new Hello3();        //第 3 个功能类实例
        hello3.show();                      //第 3 个功能类中的方法
    }
}
```

2)创建功能程序 1

在 task2 中创建功能程序 Hello1.java,编写程序代码如下:

```java
public class Hello1{
    public void show(){
        System.out.println("功能程序 1");
    }
}
```

3)创建功能程序 2

在 task2 中创建功能程序 Hello2.java,编写程序代码如下:

```java
public class Hello2{
    public void show(){
        System.out.println("功能程序 2");
    }
}
```

4)创建功能程序 3

在 task2 中创建功能程序 Hello3.java,编写程序代码如下:

```java
public class Hello3{
    public void show(){
        System.out.println("功能程序 3");
    }
}
```

5)编译多个文件的组合程序

因为在主程序中引用了各个功能,因此,主程序对各个功能程序产生了"依赖",在编译时可以先编译各个功能程序,然后再去编译主程序(也可以直接编译主程序,JDK 开发环境能直接编译多个文件的组合程序)。在 DOS 控制台中的编译过程如图 1-7 所示。

6)运行多个文件组合的程序

在控制台路径提示后面输入 java MyHello 命令后回车,则会运行 Java 联合程序,程序运行效果如图 1-8 所示。

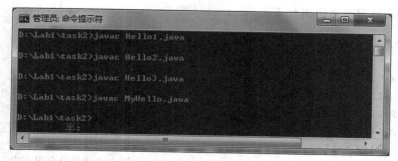

图 1-7　Java 功能文件及主文件的编译

图 1-8　联合编译并运行多个 Java 程序的运行效果

1.3.3　任务 3　使用 Eclipse 集成开发平台开发简单 Java 程序

1. 任务目的

（1）掌握在 Eclipse 集成平台中开发 Java 程序的方法。

（2）掌握在 Eclipse 集成平台中运行 Java 程序的方法。

2. 任务描述

在 Eclipse 集成平台中创建一个 Java 应用程序 Lab1，在应用程序 Lab1 中创建包 task3，在包 task3 中创建 Java 文件 HelloWorld.java，在打开的文件编辑器中编写文件，在文件编写完成后，使用 Eclipse 集成开发环境运行程序，在 Eclipse 集成平台的控制台中显示运行结果："Hello Java and Eclipse World, I'm coming!"。

3. 实验步骤

1）打开 Eclipse 程序

双击桌面或文件夹的 Eclipse 的图标，打开 Eclipse 开发平台。

2）创建 Java 项目

在 Eclipse 集成开发平台的 Package Explorer 视图中右击，在弹出的菜单中，选择 New→JavaProject 选项，如图 1-9 所示。

单击 Java Project 选项后，打开 New Java Project 对话框，填写项目名为 Lab1，选择当前系统中默认 Java SE 版本，如图 1-10 所示。

单击 Finish 按钮，完成项目的创建。

3）创建 Java 包结构

在 Package Explorer 视图中右击，在弹出的右键菜单中选择 New→Package 选项，选择

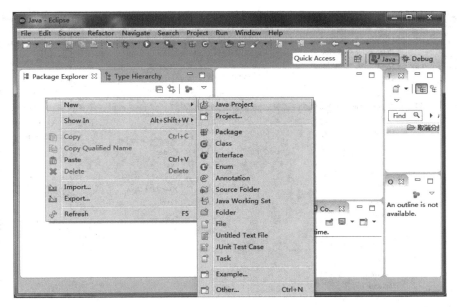

图 1-9 创建 JavaProject 项目

图 1-10 New Java Project 对话框

界面如图 1-11 所示。

单击 Package 选项,打开 New Java Package 对话框,如图 1-12 所示。

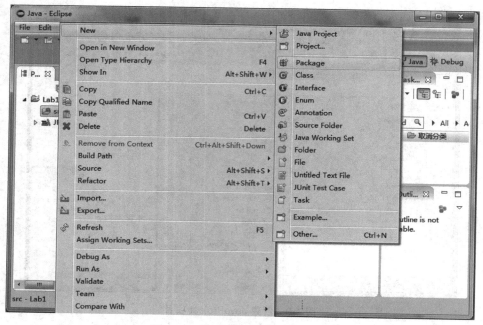

图 1-11 通过右键菜单选择创建包

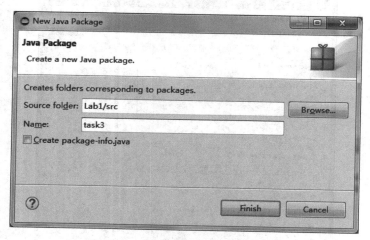

图 1-12 New Java Package 对话框

单击 Finish 按钮,可以完成包的创建,按上面方法和步骤创建包 task4。

4)创建 Java 文件

在 task3 包上右击,选择 New→Class 选项,界面如图 1-13 和图 1-14 所示。

5)编辑类文件内容

创建完成 Java 类文件后,Eclipse 会自动打开这个类文件进行编辑,修改该类的内容如下:

```
package task3;
public class HelloEclipseJava {
```

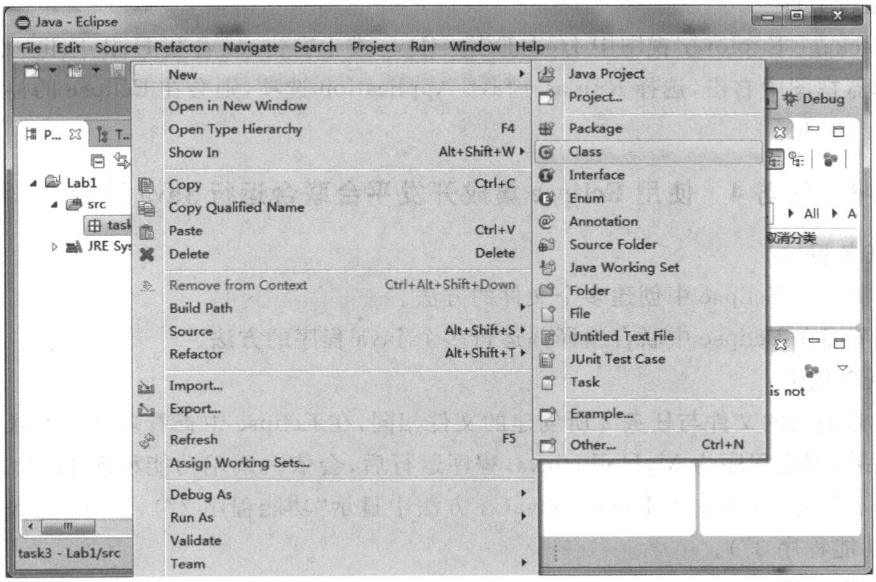

图 1-13 选择新建 Java 类文件

图 1-14 类创建界面

```
public static void main(String[] args) {
    System.out.println("Hello Java and Eclipse World,I'm coming!");
}
}
```

6）运行文件

在 Package Explorer 视图中 HelloEclipseJava 类上右击，或者在 HelloEclipseJava.java 文件编辑器视图中右击，选择 Run as→Java Application 选项，则会在 Eclipse 的控制台视图中看到运行结果。

1.3.4 任务4 使用 Eclipse 集成开发平台联合运行 Java 程序

1. 任务目的

（1）掌握在 Eclipse 中创建多个文件的方法。

（2）掌握在 Eclipse 中联合编译并运行多个 Java 程序的方法。

2. 任务描述

本任务创建的文件与任务2所创建的文件相同，在 Eclipse 中查看多个文件联合编译及运行的效果，即主程序为 MyHello.java，程序运行后，会依次调用功能程序 Hello1.java（在方法中显示"功能程序1"）、Hello2.java（在方法中显示"功能程序2"）、Hello3.java（在方法中显示"功能程序3"）。

3. 实验步骤

1）创建主程序与功能程序

按任务3的方法，创建主程序类文件 MyHello.java，功能程序 Hello1.java、Hello2.java、Hello3.java 类文件。并按任务2中各个文件的内容修改本任务的文件内容。注意保留文件顶部包的名字，主程序类文件 MyHello.java 的内容如下：

```
package task4;                                  //注意保留这里的包的名字
public class MyHello{
    public static void main(String[] args){
        Hello1 hello1=new Hello1();             //第1个功能类实例
        hello1.show();                          //第1个功能类中的方法
        Hello2 hello2=new Hello2();             //第2个功能类实例
        hello2.show();                          //第2个功能类中的方法
        Hello3 hello3=new Hello3();             //第3个功能类实例
        hello3.show();                          //第3个功能类中的方法
    }
}
```

其他的几个功能程序文件 Hello1.java、Hello2.java、Hello3.java 也要注意保留各自顶部的包的名字。

2）编译与运行多个文件

在 Eclipse 中直接运行主程序即可以完成多个文件组合的编译与运行。与任务3中运行 Java 程序的方法类似，在 Package Explorer 视图中 MyHello 类上右击，或者在 MyHello.java 文件编辑器视图中右击，选择 Run as→Java Application 选项，则会在 Eclipse 的控制台视图中看到运行结果。

4. 任务拓展

除了创建新的文件的方法，也可以将其他已经创建好的文件复制到项目中来。

1)复制原来的文件

找到原来已经编写好的文件(1个或多个),选中文件后,进行复制(或用 Ctrl+C 组合键)。

2)将复制的文件粘贴到 Eclipse 中 Java 项目的包下

选择 Java 项目中的一个包(如本任务的 task4),在包上右击,选择 Paste(或直接使用 Ctrl+V 组合键),即可以将已有的文件复制到当前的项目的包中来。

第 2 章 Java 的基本语法

2.1 实验目的

(1) 掌握基本的数据类型、了解各数值数据类型的取值范围。
(2) 掌握变量的定义与使用方法。
(3) 掌握常量的定义与使用方法。
(4) 掌握从键盘上获取输入的方法。
(5) 掌握数据类型的转换规则。
(6) 掌握表达式的使用方法。
(7) 掌握转义字符的使用方法。
(8) 掌握几类注释的使用方法。

2.2 实验任务

(1) 任务 1：编写程序显示各个数值数据类型的最值。
(2) 任务 2：声明不同类型的变量并进行赋值输出。
(3) 任务 3：根据输入圆半径的值求圆的面积。
(4) 任务 4：从键盘输入 3 个数，求它们的平均数。
(5) 任务 5：编写程序查看常用转义字符的效果。

2.3 实验内容

2.3.1 任务 1 编写程序显示各个数值数据类型的最值

1. 任务目的

(1) 掌握常用的数值数据类型。
(2) 掌握常用数值数据类型的最大值的方法。
(3) 掌握常用数值数据类型的最小值的方法。
(4) 掌握文档注释的书写方法。

2. 任务描述

编写一个 Java 程序 TypeMaxAndMinValue.java，在程序中显示常用的数值数据类型的最大值与最小值。

3. 实施步骤

1) 创建项目

在 Eclipse 中创建项目 Lab2。

2）创建包

在项目 Lab2 中创建包 task1。

3）创建文件

在 task1 包中创建 Java 文件 TypeMaxAndMinValue.java，修改该文件内容如下：

```java
package task1;        //注意这里的包名与创建的包名要一致
/**
 * 使用程序显示常用数值数据类型的最值
 * @author sf
 */
public class TypeMaxAndMinValue {
    public static void main(String[] args) {
        System.out.println("最大的 byte 值是："+Byte.MAX_VALUE);
        System.out.println("最大的 short 值是："+Short.MAX_VALUE);
        System.out.println("最大的 int 值是："+Integer.MAX_VALUE);
        System.out.println("最大的 long 值是："+Long.MAX_VALUE);
        System.out.println("最大的 float 值是："+Float.MAX_VALUE);
        System.out.println("最大的 double 值是："+Double.MAX_VALUE);
        System.out.println("最小的 byte 值是："+Byte.MIN_VALUE);
        System.out.println("最小的 short 值是："+Short.MIN_VALUE);
        System.out.println("最小的 int 值是："+Integer.MIN_VALUE);
        System.out.println("最小的 long 值是："+Long.MIN_VALUE);
        System.out.println("最小的 float 值是："+Float.MIN_VALUE);
        System.out.println("最小的 double 值是："+Double.MIN_VALUE);
    }
}
```

4）运行程序

运行程序，将会看到程序在 Eclipse 控制台中的运行结果如图 2-1 所示。

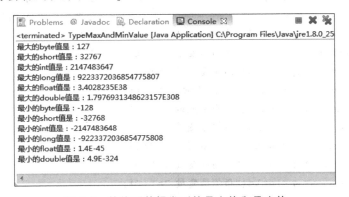

图 2-1 数值型数据类型的最大值和最小值

2.3.2 任务 2 声明不同类型的变量并进行赋值输出

1. 任务目的

（1）掌握变量的声明、赋值与使用方法。

（2）掌握数据类型变量的赋值特点。
（3）掌握行注释的使用方法。
（4）理解数据类型的隐式转换。

2. 任务描述

编写一个程序，使用常见的数据类型声明一些变量，并为这些变量赋值，然后把这些变量输出。

3. 实施步骤

1) 添加包

在项目 Lab2 中添加包 task2。

2) 添加文件并进行编辑

在包 task2 中添加 Java 类文件 DataTypes.java，修改文件的内容如下：

```java
package task2;
/**
 * 不同数据类型变量的声明、赋值与隐式转换
 * @author sf
 */
public class DataTypes {
    public static void main(String[] args) {
        byte b=0x55;                           //字节型变量声明与赋值
        short s=0x55ff;                        //短整型变量声明与赋值
        int i=1000000;                         //整型变量声明与赋值
        long l=0xffffL;                        //长整型变量声明与赋值
        char c='a';                            //字符型变量声明与赋值
        float f=0.23F;                         //单精度型变量声明与赋值
        double d=0.7E-3;                       //双精度型变量声明与赋值
        boolean B=true;                        //布尔型变量声明与赋值
        String S="这是字符串类数据类型";        //字符型变量
        System.out.println("字节型变量 b="+b);
        System.out.println("短整型变量 s="+s);
        System.out.println(" 整型变量 i="+i);
        System.out.println("长整型变量 l="+l);
        System.out.println("字符型变量 c="+c);
        System.out.println("浮点型变量 f="+f);
        System.out.println("双精度变量 d="+d);
        System.out.println("布尔型变量 B="+B);
        System.out.println("字符串类对象 S="+S);
    }
}
```

3) 运行程序

运行程序，在 Eclipse 控制台中看到程序的运行结果如图 2-2 所示。

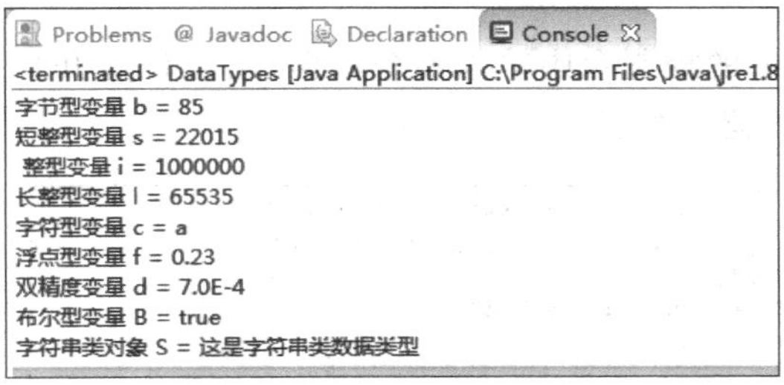

图 2-2　不同数据类型的变量赋值后的输出结果

2.3.3　任务 3　根据输入圆半径的值求圆的面积

1. 任务目的

(1) 掌握变量的声明、赋值与使用方法。
(2) 掌握常量的声明与使用方法。
(3) 掌握行注释的使用方法。
(4) 掌握获取键盘输入的方法。
(5) 掌握表达式的使用方法。

2. 任务描述

编写程序实现从键盘上输入一个半径,根据定义的常量 PI,使用表达式计算圆的半径。

3. 实施步骤

1) 添加包

在项目 Lab2 中创建包 task3。

2) 添加文件

在包 task3 中添加 Java 文件 CircleArea.java,修改该文件的内容如下:

```java
package task3;
import java.util.Scanner;
/**
 * 根据输入圆半径的值求圆的面积
 * @author sf
 */
public class CircleArea {
    private static final double PI=3.1415926;        //定义 PI 常量
    public static void main(String[] args) {
        Scanner input=new Scanner(System.in);        //输入设备
        double R=0.0,area=0.0;                       //定义圆的半径 R,圆的面积 area
        System.out.print("请输入圆的半径:");
        R=input.nextDouble();
        area=PI * R * R;                             //使用表达式计算圆的面积
        System.out.println("圆的面积为:"+area);      //输出圆的面积
        input.close();                               //关闭输入设备
    }
}
```

3）运行程序

运行程序,在 Eclipse 的控制台中输入一个数据,则会计算并显示输入的半径所对应的圆的面积。图 2-3 是一个运行实例。

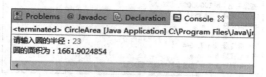

图 2-3 使用 Java 程序计算圆的面积的一个运行实例

2.3.4 任务 4 从键盘输入 3 个数并求它们的平均数

1. 任务目的

（1）掌握多个变量的声明、赋值与使用方法。
（2）掌握文档注释的使用方法。
（3）掌握行注释的使用方法。
（4）掌握表达式的使用方法。
（5）掌握键盘输入的方法。

2. 任务描述

编写一个程序,从键盘上接收 3 个双精度数,使用表达式计算出平均数后输出。

3. 实施步骤

1）添加包

在项目 Lab2 中创建包 task4。

2）添加文件

在包 task4 中添加 Java 文件 Average.java,修改该文件的内容如下：

```java
package task4;
import java.util.Scanner;

/**
 * 求从键盘上输入的 3 个数,求它们的平均数
 * @author sf
 */
public class Average {
    public static void main(String[] args) {
        Scanner input=new Scanner(System.in);      //输入设备
        double num1, num2, num3;                    //声明 3 个变量 num1、num2、num3
        double average;                             //声明平均数变量
        System.out.print("输入第 1 个数：");
        num1=input.nextDouble();
        System.out.print("输入第 2 个数：");
        num2=input.nextDouble();
        System.out.print("输入第 3 个数：");
```

```
        num3=input.nextDouble();
        //使用表达式计算平均数
        average=(num1+num2+num3)/3;
        System.out.println("您所输入的3个数据的平均数为:"+average);
        input.close();                          //关闭输入
    }
}
```

3) 运行程序

运行程序,在 Eclipse 的控制台中输入 3 个数据,则会计算并显示所输入的 3 个数据的平均数。图 2-4 是一个运行实例。

图 2-4　计算输入的 3 个数据的平均数的运行效果

2.3.5　任务 5　编写程序查看常用转义字符的效果

1. 任务目的

(1) 掌握常用转义字符的使用方法。

(2) 掌握行注释的书写方法。

2. 任务描述

编写一个 Java 程序,显示转义字符的应用效果。

3. 实施步骤

1) 添加包

在项目 Lab2 中创建包 task5。

2) 添加文件

在包 task5 中添加 Java 文件 EscapeCharacter.java,修改该文件的内容如下:

```
package task5;
/**
 * Java 转义字符应用实例
 * @author sf
 */
public class EscapeCharacter {
    public static void main(String[] args) {
        System.out.print("这是\b转义\n字符\r的测\t试实\f例,看\\看测\"试结果\'怎
                么样。");
    }
}
```

3）运行程序

运行程序,可以在 Eclipse 的控制台中看到程序的运行效果如图 2-5 所示。

图 2-5 转义字符的应用运行效果

读者根据运行显示的效果,再对照各个效果所对应的语句代码,理解各个常用的转义字符的含义与作用。

第3章 选择结构

3.1 实验目的

(1) 掌握简单 if 结构的语法与使用方法。
(2) 掌握 if-else 结构的语法与使用。
(3) 掌握多重 if-else 的语法与使用。
(4) 掌握 if-else 条件语句嵌套结构的语法与使用方法。
(5) 掌握 switch-case 结构的语法与使用方法。

3.2 实验任务

(1) 任务 1：判断键盘输入的数据是否能被 7 整除。
(2) 任务 2：成绩等级判断。
(3) 任务 3：判断键盘输入的数据是否为大写字母。
(4) 任务 4：计算销售提成。
(5) 任务 5：判断回文数。
(6) 任务 6：根据订单状态标志显示订单状态。

3.3 实验内容

3.3.1 任务 1 判断键盘输入的数据是否能被 7 整除

1. 任务目的

(1) 掌握 if-else 简单结构的使用方法。
(2) 掌握整除运算符的使用方法。
(3) 掌握()运算符的使用方法。

2. 任务描述

对任意一个从键盘上输入的整数，判断其是否能被 7 整除，如果能被 7 整除输出该数除以 7 的商，否则，输出信息"不能被 7 整除"。

3. 实施步骤

1) 创建 Java 项目

在 Eclipse 中创建 Java 项目 Lab3。

2) 创建包

在项目 Lab3 中创建包 task1。

3）创建 Java 文件并进行编辑

在包 task1 中创建文件 DivBySeven.java，修改该文件的内容如下：

```java
package task1;
import java.util.Scanner;
/**
 * 从键盘上输入一个数,检查是否能被7整除
 * @author sf
 */
public class DivBySeven {
    public static void main(String[] args) {
        Scanner input=new Scanner(System.in);        //实例化输入
        int iNum=0;                                  //声明变量并赋初值
        System.out.print("请输入一个数：");
        iNum=input.nextInt();
        if(iNum %7==0){
            System.out.println("该数除以7的商为："+(iNum/7));
        }else{
            System.out.println("不能被7整除。");
        }
        input.close();                               //关闭输入
    }
}
```

运行程序,输入不同的整数数据,观察运行结果。

4. 任务拓展

1）奇偶数判断

任务描述：输入一个数,如果这个数在除以 2 后的余数为 0,则屏幕上就会显示 The number is Even,否则就在屏幕上显示 The number is Odd。完整代码如下：

```java
package task1;
import java.util.Scanner;
/**
 * 奇偶数判断
 * @author sf
 */
public class EvenOrOdd {
    public static void main(String[] args) {
        Scanner input=new Scanner(System.in);        //实例化输入
        System.out.print("请输入一个整数：");
        int iNum=input.nextInt();                    //获取一个输入
        if(iNum%2==0){
            System.out.println("The number is Even。");
        }else{
            System.out.println("The number is Odd。");
        }
```

```
            input.close();                              //关闭输入
    }
}
```

运行后输入数据测试,观察执行结果。

2) 选项判断。

任务描述:从键盘输入一个数字,如果这个数在 1~3 之间,将在屏幕上显示这个数字,否则将显示 Invalid Choice。完整代码如下:

```
package task1;
import java.util.Scanner;
/**
 * 选项判断
 * @author sf
 */
public class OptionJudge {
    public static void main(String[] args) {
        Scanner input=new Scanner(System.in);        //实例化输入
        int iNum=0;                                  //存储选定的选择项的变量
        System.out.print("请输入一个选择项的值,范围为[1,3]: ");
        iNum=input.nextInt();                        //接收输入
        if(iNum>=1 && iNum<=3){
            System.out.println("选择的项为:"+iNum);
        }else{
            System.out.println("Invalid Choice");
        }
        input.close();                               //关闭输入
    }
}
```

运行后输入数据测试,观察执行结果。

3.3.2 任务 2 成绩等级判断

1. 任务目的

(1) 掌握 if-else 多分支结构的使用方法。
(2) 掌握键盘输入内容的方法。

2. 任务描述

从键盘上输入一个学生的成绩,如果这个成绩在[90,100]内,则输出"优秀";如果这个成绩在[70,90)内,则输出"良好";如果这个成绩在[60,70)内,则输出"及格";低于 60 分时输出"不及格",高于 100 分时,输出"输入的内容不合法"。

3. 实施步骤

1) 创建包

在项目 Lab3 中创建包 task2。

2）创建并修改文件

在 task2 包中创建文件 ScoreGrade.java，修改该文件的内容如下：

```java
package task2;
import java.util.Scanner;
/**
 * 成绩等级
 * @author sf
 */
public class ScoreGrade {
    public static void main(String[] args) {
        Scanner input=new Scanner(System.in);        //实例化输入设备
        int score=0;                                 //成功变量
        System.out.print("请输入一个成绩的值：");
        score=input.nextInt();
        if(score>100){
            System.out.println("输入的数据不合法。");
        }else if(score>=90){
            System.out.println("优秀");
        }else if(score>=70){
            System.err.println("良好");
        }else if(score>=60){
            System.out.println("及格");
        }else{
            System.out.println("不及格");
        }
        
        input.close();                               //关闭输入
    }
}
```

3）运行程序

运行该程序，输入不同的数据（特别是边界值），查看运行效果，如下数据为不同测试用例信息：

```
请输入一个成绩的值：101
输入的数据不合法。
请输入一个成绩的值：100
优秀
请输入一个成绩的值：70
良好
请输入一个成绩的值：60
及格
请输入一个成绩的值：59
不及格
```

3.3.3 任务3 判断键盘输入的数据是否为大写字母

1. 任务目的
(1) 掌握键盘输入字符串的方法。
(2) 掌握字符串转换为字符数组的方法。
(3) 掌握大写字母与小写字母的 ASCII 码值。
(4) 掌握大写字母与小写字母的 ASCII 码的差值。

2. 任务描述
从键盘输入一个字符串,判断输入的字符串是否是含有大写字母,如果是大写字母将其转换成小写字母,否则不用改变,修改完成后,将输入的内容输出。

3. 任务分析
根据任务描述,要完成这个任务,需要检查字符串中的每个字符是否为大写字母,而判断一个字符是否为大写字母,就要用到字符的 ASCII 值,因此要完成本任务,就要将输入的字符串转换为字符数组,然后判断每个字符的 ASCII 值,这里要使用大写字母与小写字母的 ASCII 码值进行转换,大写字母的 ASCII 码值如下:

A	B	C	D	E	F	G	H	I	J	K	L	M	N	O	P	Q	R	S	T	U	V	W	X	Y	Z
65	66	67	68	69	70	71	72	73	74	75	76	77	78	79	80	81	82	83	84	85	86	87	88	89	90

小写字母的 ASCII 码值如下:

a	b	c	d	e	f	g	h	i	j	k	l	m	n	o	p	q	r	s	t	u	v	w	x	y	z
97	98	99	100	101	102	103	104	105	106	107	108	109	110	111	112	113	114	115	116	117	118	119	120	121	122

由大写字母与小写字母的 ASCII 码表,可以看出,大写字母的 ASCII 范围为[65,90],小写字母的 ASCII 范围为[97,122],小写字母比对应大写字母的 ASCII 的值大 32。由此可得出程序的设计思路。
(1) 从键盘输入一个字符串。
(2) 将字符串转换为字符数组。
(3) 检查每个字符,若是大写字母,则转换为小写字母输出,否则不转换直接输出。

4. 实施步骤
1) 创建包
在项目 Lab3 中创建包 task3。
2) 创建文件并修改
在包 task3 中创建 Java 类文件 Big2SmallLetter.java,修改文件的内容如下:

```
package task3;
import java.util.Scanner;

/**
 * 输入一个字母,如果是大写字母,将其转换为小写输出,否则直接输出  *
```

```
 @author sf
 */
public class Big2SmallLetter {
    public static void main(String[] args) {
        Scanner input=new Scanner(System.in);      //初始化输入
        String str;                                 //字符串变量
        System.out.print("请输入一个字符串：");
        str=input.next();
        char[] myCharArr=str.toCharArray();
        for(int i=0; i<myCharArr.length; i++){
            if(myCharArr[i]>=65 && myCharArr[i]<=90){
                myCharArr[i]+=32;
            }
            System.out.print(myCharArr[i]);
        }
        input.close();                              //关闭输入
    }
}
```

3）运行程序

图 3-1 所示为该程序的一个运行实例。

图 3-1 大写字母转换为小写字母的一个实例

3.3.4 任务4 计算销售提成

1. 任务目的

（1）掌握 if-else 单分支结构的使用方法。
（2）掌握 if-else 多分支结构的使用方法。
（3）掌握从键盘输入数据的方法。

2. 任务描述

公司对月销售额大于等于￥10000 的销售人员付给其 10% 的提成。提成在月末计算。编写程序，以便根据销售人员的销售额计算其应得的提成。

3. 实施步骤

1）创建包

在项目 Lab3 中创建包 task4。

2）创建文件并修改

在包 task4 中创建 Java 类文件 SalesCommissions.java，修改文件的内容如下：

```
package task4;
```

```java
import java.util.Scanner;
/**
 * 计算销售提成
 * @author sf */
public class SalesCommissions {
    public static void main(String[] args) {
        Scanner input=new Scanner(System.in);          //实例化输入
        System.out.print("请输入销售额(单位：￥)：");
        float sale=input.nextFloat();                  //销售额
        float royalty=0.0f;                            //提成
        if(sale>10000){
            royalty=sale * 0.1f;
        }
        System.out.println("销售提成为：￥"+royalty);
        input.close();
    }
}
```

3）运行程序

图 3-2 所示为该程序的两个运行实例。

(a)　　　　　　　　　(b)

图 3-2　销售提成运行情况

4. 任务拓展

为了保证社会正常、公平、稳定地运行,国家对于高工资人群实施税收政策,该政策规定,月薪 3000 元以下不收税,月薪(3000,5000]内的超过 3000 的部分收 5％的税；月薪(5000,10000]内的超过 5000 的部分收 10％的税,超过 3000 的部分收 5％的税；月薪(10000,20000]内的超过 10000 的部分收 15％的税,超过 5000 的部分收 10％的税,超过 3000 的部分收 5％的税；对于工资高于 20000 元的月薪,高于 20000 的部分收 25％的税,其他超过的部分按前面的标准收税,请编写这个工资收税程序。

1）任务分析

根据任务描述,可以看出,税金是进行累加的,而当收入超过某个值时,对于该值前面的固定区间,如对于 5100,它前面 3000～5000 的税金是固定的,因此可以将这些固定区间的固定税金先计算出来。

2）创建文件并进行修改

在包 task4 中创建 Java 文件 CalcRevenue.java,修改该文件的内容如下：

```java
package task4;
import java.util.Scanner;
```

```java
/**
 * 根据收入计算税金
 * @author sf */
public class CalcRevenue {
    public static void main(String[] args) {
        Scanner input=new Scanner(System.in);
        System.out.print("请输入工资收入(单位：￥)：");
        float salary=input.nextFloat();               //工资收入
        float revenue=0.0f;                           //税金
        //3000~5000间的固定税金100
        float rev3kto5k=(5000-3000)*0.05f;
        //5000~10000间的固定税金500
        float rev5kto1w=(10000-5000)*0.1f;
        //10000~20000间的固定税金1500
        float rev1wto2w=(20000-10000)*0.15f;
        if(salary<=3000){
            revenue=0.0f;
        }else if(salary<=5000){
            revenue+=(salary-3000)*0.05f;
        }else if(salary<=10000){
            revenue+=(salary-5000)*0.1f+rev3kto5k;
        }else if(salary<=20000){
            revenue+=(salary-10000)*0.15f+rev3kto5k+rev5kto1w;
        }else{
            revenue+=(salary-20000)*0.25f+rev3kto5k+rev5kto1w+rev1wto2w;
        }
        System.out.println("该收入要征收的税金为："+revenue);
        input.close();
    }
}
```

3）运行程序

该程序的几个运行实例如下：

请输入工资收入(单位：￥)：3000
该收入要征收的税金为：￥0.0
请输入工资收入(单位：￥)：4500
该收入要征收的税金为：￥75.0
请输入工资收入(单位：￥)：7000
该收入要征收的税金为：￥300.0
请输入工资收入(单位：￥)：12000
该收入要征收的税金为：￥900.0
请输入工资收入(单位：￥)：21000
该收入要征收的税金为：￥2350.0

3.3.5 任务5 判断回文数

1. 任务目的

（1）掌握if-else分支结构的使用方法。

(2) 掌握整型数据转换为字符串数组的方法。
(3) 掌握使用键盘输入数据的方法。
(4) 掌握嵌套的 if-else 分支结构的使用方法。

2. 任务描述

回文数是指该数所含有的数字逆序排序后,得到的新的数字与原来的数字相同,如 12121、3223 都是回文数。编写一个 Java 应用程序,判断从键盘上输入一个整数是否为回文数,并将这个数据和判断结果输出。

3. 实施步骤

1) 创建包

在项目 Lab3 中创建包 task5。

2) 创建文件并修改

在包 task5 中创建 Java 类文件 Palindrome.java,修改文件的内容如下:

```java
package task5;
import java.util.Scanner;
/**
 * 回文数判断
 * @author sf */
public class Palindrome {
    public static void main(String[] args) {
        Scanner input=new Scanner(System.in);        //输入实例
        System.out.print("请输入一个整型数字:");
        long num=input.nextLong();                   //输入一个整数
        String str=num+"";                           //将数字转换为字符串
        char[] arr=str.toCharArray();                //将字符串转换为字符数组
        int len=arr.length;                          //字符串的长度
        boolean flag=true;                           //回文数标志,默认为是
        //在循环中判断两端的字符是否相同
        for(int i=0;i<len/2;i++){
            //如果两端对应的字符不同,则不是回文数,结束循环
            if(arr[i]!=arr[len-1-i]){
                flag=false;                          //将回文数标志为否
                break;                               //结束循环
            }
        }
        //根据标志,输出前面输入的数据是否为回文数
        if(flag){
            System.out.println("您所输入的数据:"+num+",为回文数。");
        }else{
            System.out.println("您所输入的数据:"+num+",不是回文数。");
        }
        input.close();                               //关闭输入
    }
}
```

3) 运行程序

该程序的几个运行实例如下：

请输入一个整型数字：4521
您所输入的数据：4521,不是回文数。
请输入一个整型数字：1343431
您所输入的数据：1343431,为回文数。

4. 任务拓展

将某个指定范围内的回文数全部输出来,如将 10000 以内的回文数输出来。

1) 任务分析

该任务是要对指定的一个范围内的所有的数据进行检查,检查每个数据是否为回文数,因此可以将回文数的检查功能提取出来作为一个独立的方法,在循环中反复地调用这个方法进行检查并输出。

2) 创建文件并修改

在包 task5 中创建 Java 类文件 PalindromeIn1W.java,修改文件的内容如下：

```java
package task5;
/**
 * 检查 10000 以内的所有回文数,并进行输出
 * @author sf
 */
public class PalindromeIn1W {
    public static void main(String[] args) {
        //在循环中检查 10000 以内的每个数据是否为回文数
        String str="";                    //构建要显示的所有的回文数所组成字符串
        int count=0 ;                     //回文数的计数
        for(int i=0;i<=10000;i++){
            boolean flag=isPalindrome(i);   //检查每个数据是否为回文数
            //根据标志,输出前面输入的数据是否为回文数
            if(flag){
                count++;                  //计数器计数
                //字符串是否为空串,为空时,添加第 1 个元素
                if("".equals(str)){
                    str=i+"";
                }else{                    //字符串不为空时,添加第 2 个~第 n 个元素
                    str+=","+i ;
                }
            }
        }
        //显示 10000 以内所有的回文数信息
        System.out.println("10000 以内的回文数有["+count+"]个,各个回文数如下：");
        System.out.println(str);
    }
```

```java
/**
 * 判断一个数据是否为回文数的检查方法
 * @param num 要检查的数据
 */
public static boolean isPalindrome(int num){
    String str=num+"";                    //将数字转换为字符串
    char[] arr=str.toCharArray();         //将字符串转换为字符数组
    int len=arr.length;                   //字符串的长度
    boolean flag=true;                    //回文数标志,默认为是

    //在循环中判断两端的字符是否相同
    for(int i=0;i<len/2;i++){
        //如果两端对应的字符不同,则不是回文数,结束循环
        if(arr[i]!=arr[len-1-i]){
            flag=false;                   //将回文数标志为否
            break;                        //结束循环
        }
    }
    return flag;
}
```

3) 运行程序

程序的运行效果如图 3-3 所示。

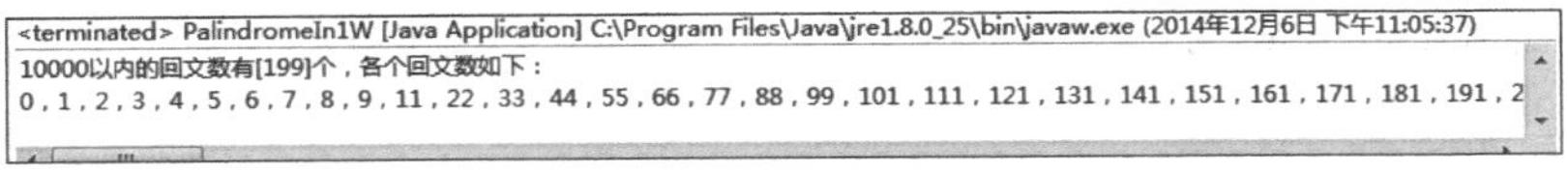

图 3-3 指定范围内的所有回文数程序的运行效果

3.3.6 任务 6 根据订单状态标志显示订单状态

1. 任务目的

(1) 掌握 switch-case 开关分支结构的使用方法。
(2) 掌握从键盘输入数据的方法。

2. 任务描述

在电子商务系统中,为了明确地表现订单的状态,方便买家和卖家的查询及处理,一般给定订单的几个状态标识,在显示时,要将这些标识转换为明确的提示信息,如在一个电子商务平台系统中规定如下订单标识及含义。

0:订单取消。
10:新订单,未付款。
20:已付款,未发货。
30:已发货,未收货。
40:已收货,未评价。

50：已评价。

请编写一个程序，根据对应的标识值，输出显示提示信息。

3. 任务分析

根据任务描述信息，可知可以将标识与提示信息作为一个独立的模块（这里使用一个方法来实现），因为标识值为常量，因此采用 switch-case 开关分支结构来实现这个程序。

4. 实施步骤

1) 创建包

在项目 Lab3 中创建包 task6。

2) 创建文件并修改

在包 task6 中创建 Java 类文件 OrdersState.java，修改文件的内容如下。

```java
package task6;
import java.util.Scanner;
/**
 * 订单状态提示信息
 * @author sf
 */
public class OrdersState {
    public static void main(String[] args) {
        Scanner input=new Scanner(System.in);    //输入实例
        System.out.print("请输入一个状态标识：");
        int order_state=input.nextInt();
System.out.println("当前的订单状态为："+getOrderState(order_state));
        input.close();                                   //关闭输入
    }
    /**
     * 将订单状态标识转换成对应提示信息
     * @param order_state 订单标志
     * @return 订单状态提示信息
     */
    public static String getOrderState(int order_state){
        String str="";
        switch (order_state) {
        case 10:
            str="新订单,未付款";
            break;
        case 20:
            str="已付款,未发货";
            break;
        case 30:
            str="已发货,未收货";
            break;
        case 40:
            str="已收货,未评价";
```

```
            break;
        case 50:
            str="已评价";
            break;
        default:
            str="已取消";
            break;
        }
        return str;
    }
}
```

3）运行程序

该程序的一个运行实例如下：

请输入一个状态标识：20
当前的订单状态为：已付款,未发货

第4章 循环结构

4.1 实验目的

(1) 掌握解决重复问题的3种结构。
(2) 掌握 for 循环的语法与使用。
(3) 掌握 while 循环的语法与使用。
(4) 掌握 do-while 循环的语法与使用。
(5) 理解3种循环各自应用的场合。
(6) 掌握 break 和 continue 关键字的使用。
(7) 理解并掌握循环嵌套的特点和使用。

4.2 实验任务

(1) 任务1:摄氏温度到华氏温度的转换表。
(2) 任务2:抽奖。
(3) 任务3:求和。
(4) 任务4:break 和 continue。
(5) 任务5:猜数字游戏。
(6) 任务6:马克思手稿中的数学题。

4.3 实验内容

4.3.1 任务1 摄氏温度到华氏温度的转换表

1. 任务目的

(1) 掌握 for 语句的语法。
(2) 能够灵活运用 for 语句解决重复问题。

2. 任务描述

使用 for 语句按 5℃ 的增量打印出一个从摄氏温度到华氏温度的转换表,摄氏温度到华氏温度的转换公式为 h=c*9/5+32。

3. 实施步骤

1) 算法分析

根据任务描述可知,c 从 0 开始要重复地执行语句 h=c*9/5+32,显然要借助于循环结构来解决。使用循环结构要搞清楚:

循环结束的条件是什么?

循环体是什么?

2) 参考代码

```java
public class Task1 {
    public static void main(String[] args) {
        int h, c;
        System.out.println("摄氏温度 华氏温度");
        for (c=0; c<=40; c+=5){
            h=c * 9/5+32;
            System.out.println( c+"\t"+h);
        }
    }
}
```

3) 运行程序

观察结果,如图 4-1 所示。

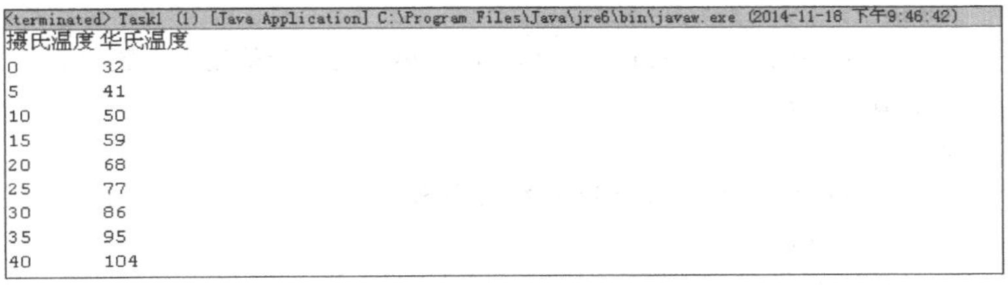

图 4-1 摄氏温度到华氏温度转换表

4. 任务拓展

(1) 使用 while 循环语句改写程序。

(2) 使用 do-while 循环语句改写程序。

4.3.2 任务 2 抽奖

1. 任务目的

(1) 掌握 while 语句的语法。

(2) 能够灵活运用 while 语句解决重复问题。

2. 任务描述

从键盘输入数字 1、2、3 后,可显示抽奖得到的奖品,如果为 1,输出"恭喜你得大奖,一辆汽车!";如果为 2,输出"不错呀,你得到一台笔记本电脑!";如果为 3,输出"没有白来,你得到一台冰箱!",如果输入其他数字显示"真不幸,你没有奖品!下次再来吧";如果输入 0,则停止抽奖。

3. 实施步骤

1) 算法分析

Step1:从键盘输入数字。

Step2:如果数字不是 1、2、3,则显示"真不幸,你没有奖品!下次再来吧",程序终止,否则进入 Step3。

Step3：根据输入数字的不同,显示不同的奖品,执行Step1。

2) 参考代码

```java
import java.util.Scanner;
public class Task2 {
    public static void main(String[] args) {
        Scanner input=new Scanner(System.in);
        System.out.println("请输入数字(1、2、3),输入0停止抽奖");
        int number=1;
        while (number!=0) {
            System.out.println("请输入数字(1、2、3)");
            number=input.nextInt();
            switch (number) {
            case 1:
                System.out.println("恭喜你得大奖,一辆汽车!");
                break;
            case 2:
                System.out.println("不错呀,你得到一台笔记本电脑!");
                break;
            case 3:
                System.out.println("没有白来,你得到一台冰箱!");
                break;
            default:
                System.out.println("真不幸,你没有奖品!下次再来吧");
            }
        }
    }
}
```

3) 运行程序

观察结果,如图4-2所示。

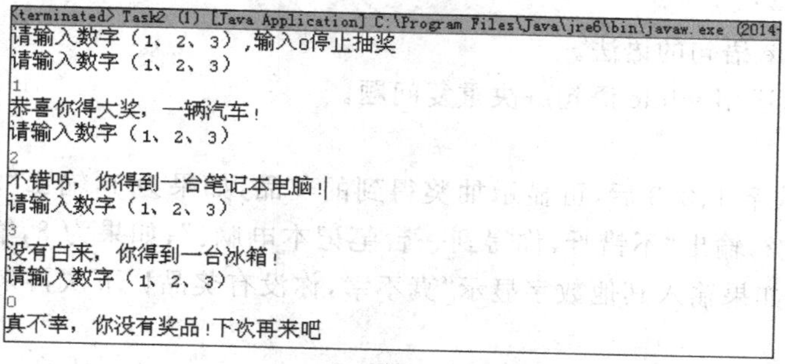

图4-2 抽奖程序运行结果

4. 任务拓展

(1) 使用for循环语句改写程序。

(2) 使用do-while循环语句改写程序。

4.3.3 任务3 求和

1. 任务目的

(1) 掌握 do-while 语句的语法。

(2) 能够灵活运用 do-while 语句解决重复问题。

2. 任务描述

求 1+2+…+100 之和,并将求和表达式与所求的和显示出来。

3. 实施步骤

1) 算法分析

根据任务描述,这是一个典型的累加和问题。

2) 参考代码

```
public class Task3 {
    public static void main(String[] args) {
        int sum=0;
        int i=1;
        do {
            sum=sum+i;
            i++;
        } while (i<=100);
        System.out.println("sum="+sum);
    }
}
```

3) 运行程序

观察结果,如图 4-3 所示。

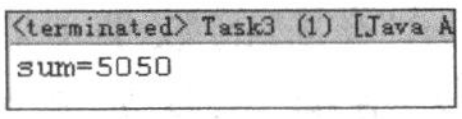

图 4-3 求和

4. 任务拓展

(1) 使用 for 循环语句改写程序。

(2) 使用 while 循环语句改写程序。

4.3.4 任务4 break 和 continue

1. 任务目的

(1) 理解 break 关键字的特点。

(2) 理解 continue 关键字的特点。

(3) 能够灵活运用 break 和 continue 关键字。

2. 任务描述

阅读代码,体会 break 和 continue 关键字的特点和区别。

3. 实施步骤

(1) 阅读如下程序代码段,输出结果是什么?

```
int sum=0;
for( int i=0;i<5;i++){
```

```
        if(i==3) {
            break;
        }
        sum=sum+i;
    }
    System.out.println("sum="+sum);
```

（2）在 VC 环境下运行该程序，比较输出结果和你分析得出的结果是否一致？

（3）将 break 替换为 continue 结果是什么？

（4）讨论：break 和 continue 关键字的特点。

4. 程序拓展

修改上述程序，不使用 break 关键字达到同样的效果。

4.3.5 任务 5 猜数字游戏

1. 任务目的

（1）理解并掌握 3 种循环结构的适用场合。

（2）能够灵活地运用循环结构解决实际问题。

2. 任务描述

计算机随机产生一个 1~100 之间的数，然后我们去猜，我们有可能不是猜大了就是猜小了，如果在猜的过程中计算机又不给我们些暗示，我想要在 1~100 之间的数中猜出到底计算机心里想的哪个数，真的是太难了。还好计算机是比较通情达理的，如果猜大了，它会提示"你猜的数大了，继续猜吧！"，这样下次猜的时候就会往小的方向猜；如果猜小了，它会提示"你猜的数小了，继续猜吧！"，再猜的时候可以往大的方向猜，这样一来就会逐步地缩小猜测范围，最终如果在次数没有任何限制的情况下，一定会猜到计算机心里想的那个数。最后还可以根据猜的次数，计算机给出相应的信息，如果 1 次就猜对了，输出"快来看，上帝…"；如果猜的次数在 2~6 次之间，则输出"这么快就猜对了，你很聪明啊！"；如果猜的次数超过 6 次，则输出"猜了半天才猜出来，小同志，尚须努力啊！"；如果限制次数的话，到最后可能还没有猜出计算机心里所想，游戏就提示你 Game Over 了等。

3. 实施步骤

1）算法分析

猜数字游戏的过程是个重复的过程，需要使用循环结构。究竟该选择 for 循环结构、while 循环结构还是 do-while 循环结构呢？再来回顾一下这 3 种结构的特点：for 循环适合解决一开始就能确定循环次数的情境，while 循环当满足一定条件时就会执行循环操作，do-while 循环和 while 循环的区别是至少会执行一遍循环操作。玩猜数字游戏再聪明的人也至少需要猜一次，循环操作至少会执行一遍，所以优先选用 do-while 循环。

循环条件应该是什么呢？大家想想游戏什么时候才算结束呢？那当然是用户猜对了。如何知道用户是否猜对了呢？那当然是用户猜的数和随机产生的那个数相等的时候就不需要再猜了。所以循环条件应该是用户没有猜对时，即用户猜的数和随机产生的那个数如果不相等，用户就需要继续猜。

解决这个问题大致需要以下 4 个步骤。

(1) 随机产生一个 1～100 之间的随机数并声明一个变量 randNumber 保存这个数;声明一个变量 count 初始值为 0,记录用户猜的次数。

(2) 声明一个变量 guess 保存用户输入的猜测并与 randNumber 进行比较,如果 guess 小于 randNumber,提示用户"你猜的数太小了,继续猜吧!";如果 guess 大于 randNumber,提示用户"你猜的数太大了,继续猜吧!";使 count 的值增 1。

(3) 如果 guess 不等于 randNumber,则回到(2)继续;否则执行(4)。

(4) 根据用户猜测的次数及 count 的值,输出相应的信息。

2) 根据算法分析,编写程序

参考代码如下:

```java
import java.util.Random;
import java.util.Scanner;
public class Task5 {
    public static void main(String[] args) {
        int randNumber;              //定义存放产生随机数的变量
        int guess;                   //存放用户所猜的数
        int count=0;                 //统计用户所猜测的次数
        //产生随机数
        Random rand=new Random();
        randNumber=rand.nextInt(100)+1;
        //输入用户所猜的数,直到猜对为止,并统计用户所猜的次数
        do {
            System.out.println("请输入你猜的数:");
            Scanner input=new Scanner(System.in);
            guess=input.nextInt();
            if (guess>randNumber)
                System.out.println("你猜的数太大了,继续猜吧!");
            else if (guess<randNumber)
                System.out.println("你猜的数太小了,继续猜吧!");
            count++;
        } while (guess!=randNumber);
        //根据次数打印出不同的信息
        switch (count) {
            case 1:
                System.out.println("快来看,上帝…");
                break;
            case 2:
            case 3:
            case 4:
            case 5:
            case 6:
                System.out.println("这么快就猜对了,你很聪明啊!");
                break;
            default:
                System.out.println("猜了半天才猜出来,小同志,尚须努力啊!");
                break;
```

```
        }
    }
}
```

3) 运行程序

观察结果,如图 4-4 所示。

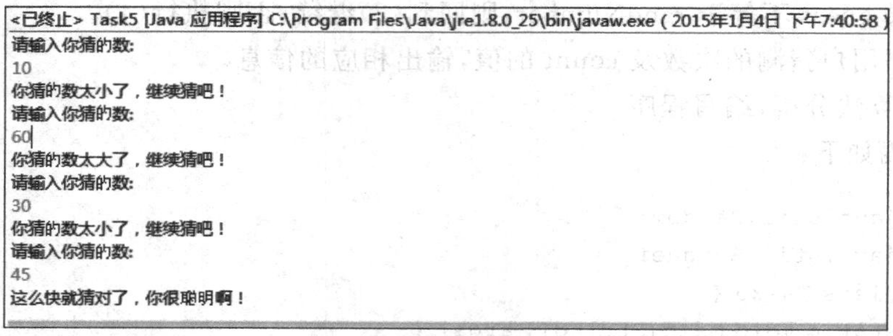

```
<已终止> Task5 [Java 应用程序] C:\Program Files\Java\jre1.8.0_25\bin\javaw.exe (2015年1月4日 下午7:40:58)
请输入你猜的数:
10
你猜的数太小了,继续猜吧!
请输入你猜的数:
60
你猜的数太大了,继续猜吧!
请输入你猜的数:
30
你猜的数太小了,继续猜吧!
请输入你猜的数:
45
这么快就猜对了,你很聪明啊!
```

图 4-4 猜数字游戏过程

4. 程序拓展

如果玩了一次还不过瘾,还想让计算机重新生成一个随机数继续猜,请修改程序。

4.3.6 任务6 马克思手稿中的数学题

1. 任务目的

(1) 理解并掌握循环嵌套结构。
(2) 能够灵活地运用循环嵌套解决实际问题。
(3) 能够灵活运用穷举法解决实际问题。

2. 任务描述

编程求解马克思手稿中的数学题。马克思手稿中有一道趣味数学题:有 30 个人,其中有男人、女人和小孩,在一家饭馆里吃饭共花了 50 先令,每个男人各花 3 先令,每个女人各花 2 先令,每个小孩各花 1 先令,问男人、女人和小孩各几人?

3. 实施步骤

1) 算法分析

设男人个数为 x,女人个数为 y,小孩个数为 z,则可以得到以下方程组:

$$x+y+z=30$$
$$3*x+2*y+z=50$$

两个方程组,3 个未知数,通过我们所掌握的数学知识,这个解是无法直接求解出来的。但是通过上述方程组,可以初步的确定 x、y 的最大取值:x 的最大取值为 16,y 的最大取值为 25,z 的值 $=30-x-y$。

x	y	z=30−x−y	3*x+2*y+z 是否等于 50
1	1	28	否
1	2	27	否
1	3	26	否
⋮	⋮	⋮	⋮

1	25	4	否
2	1	27	否
2	2	26	否
⋮	⋮	⋮	⋮
2	16	12	是
2	3	25	否
⋮	⋮	⋮	⋮
2	25	3	否
…	…	…	…

通过以上分析,可知:在男人人数为1的情况下,女人要依次从1试探到25,女人试探了一遍,男人才开始下一轮。这个题目又具备了嵌套循环的典型特征。

2)根据算法分析,编写程序

参考代码如下:

```java
public class Task6 {
    public static void main(String[] args) {
        for(int x=1;x<=16;x++){            //外循环试探男人的个数
            for(int y=1;y<=25;y++){        //内循环试探女人的个数
                int z=30-x-y;              //x,y确定了,z的值可以直接取得
                //此时只需要判断另外一个方程式是否成立
                if(3*x+2*y+z==50){
                    System.out.println("x="+x+"y="+y+"z="+z);
                }
            }
        }
    }
}
```

3)运行程序

观察结果,如图4-5所示。

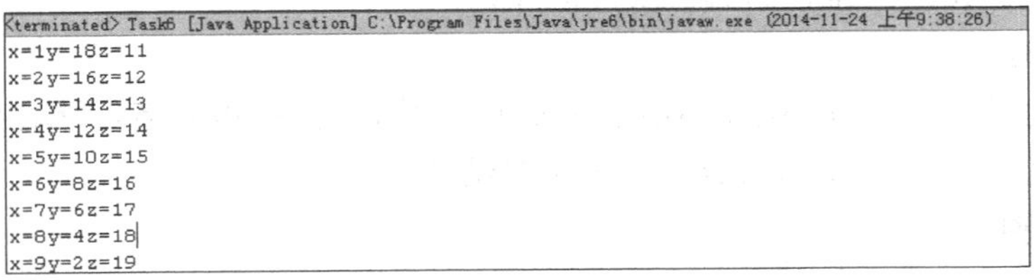

```
<terminated> Task6 [Java Application] C:\Program Files\Java\jre6\bin\javaw.exe (2014-11-24 上午9:38:26)
x=1y=18z=11
x=2y=16z=12
x=3y=14z=13
x=4y=12z=14
x=5y=10z=15
x=6y=8z=16
x=7y=6z=17
x=8y=4z=18
x=9y=2z=19
```

图4-5 马克思手稿中的数学题求解

4. 程序拓展

编程实现百钱买百鸡问题:100元钱买100只鸡,公鸡每只五元,母鸡每只三元,小鸡三只一元钱,问公鸡、母鸡、小鸡各买多少只?

第5章 数　　组

5.1　实验目的

（1）理解并掌握数组的声明、创建、初始化和数组的遍历。
（2）理解数组创建时的内存变化情况。
（3）掌握经典的排序算法——冒泡排序和选择法排序。
（4）掌握经典的二分查找的算法思路。
（5）掌握 Arrays 类的使用。
（6）掌握对象数组的创建和使用。
（7）掌握二维数组的创建和使用。

5.2　实验任务

（1）任务1：成绩统计。
（2）任务2：食堂饭菜质量评价。
（3）任务3：打印杨辉三角形。

5.3　实验内容

5.3.1　任务1　成绩统计

1. 任务目的

（1）数组的声明、创建、初始化和数组的遍历。
（2）能够灵活运用一维数组解决实际问题。

2. 任务描述

从键盘上输入若干学生(假设不超过100)的成绩,计算平均成绩,并输出高于平均分的学生人数及成绩。这里约定输入成绩为101时结束。

3. 实施步骤

1) 算法分析

首先简要分析一下求平均分的算法思路。

step1：定义数组,数组的长度为100。

step2：循环录入学生成绩并累加,如果录入成绩为101,则跳出循环。

step3：平均分＝累加和/录入成绩的个数。

下面再简要分析一下输出高于平均分的学生人数及成绩的实现思路：设置一计数器,初始值为0,遍历数组,发现高于平均分的就输出并对计数器加1。

2）参考代码

```java
public class Task1 {
    public static void main(String[] args) {
        float[] score=new float[100];
        Scanner input=new Scanner(System.in);
        float sum=0;                     //累加和
        int i;
        for(i=0;i<score.length;i++){
            System.out.print("请输入第"+(i+1)+"名的学生成绩：");
            score[i]=input.nextFloat();
            if(score[i]==101){
                break;
            }
            sum+=score[i];
        }
        float average=sum/i;
        System.out.println("平均分为："+average);
        int count=0;                     //统计高于平均分的学生人数
        for(int j=0;j<i;j++){
            if(score[j]>average){
                System.out.println("第"+(j+1)+"名的学生成绩为："+score[j]);
                count++;
            }
        }
        System.out.println("成绩高于平均分的有"+count+"人");
    }
}
```

3）运行程序

观察结果，如图 5-1 所示。

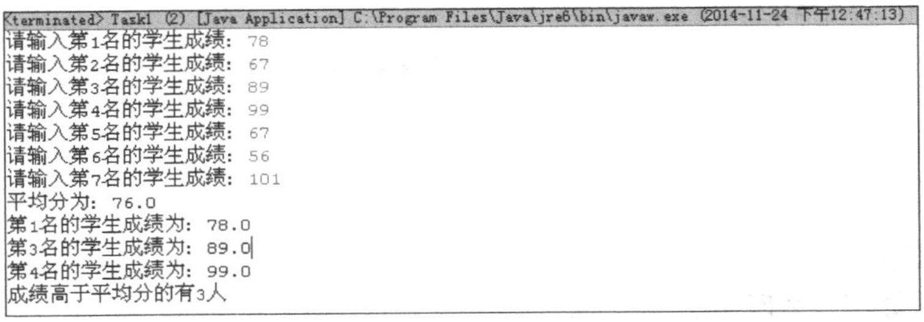

图 5-1　成绩统计

4. 任务拓展

（1）输出最高分和最低分。

（2）成绩排序输出。

5.3.2 任务 2 食堂饭菜质量评价

1. 任务目的

能够灵活运用数组解决实际问题。

2. 任务描述

要求 20 名同学对学生食堂饭菜的质量进行 1~5 的评价(1 表示很差,5 表示很好)。将这 20 个结果输入整型数组,并对打分结果进行分析。

3. 实施步骤

1)算法分析

我们希望统计出每个分数对应的学生人数。学生最终的打分情况可以借助于一个整型数组 answers 来保存。首先定义一个包含 20 个打分结果的 answers 数组,然后再定义一个包含 6 个元素的数组 frequency 来统计各种评价的次数,frequency 中的每个元素此时都被看成了一个得分的计数器,其默认的初始值为 0。在这里为何要将数组长度定义为 6 呢?我们想让 frequency[1]统计的是分值 1 的次数,frequency[2]统计的是分值 2 的次数,…,frequency[5]统计的是分值 5 的次数,这样正好一一对应起来,这里忽略掉 frequency[0]。比如读入第一个学生的评价,他的评价是 5 分,就将 frequency[5]的计数值加 1。当遍历完 answers 数组后,对应的 frequency 数组里面也已经统计完了。

2)参考代码

```java
public class Task2 {
    public static void main(String[] args) {
        int[] answers={3,1,2,5,4,2,2,3,4,5,1,2,3,4,2,1,3,2,
            4,2};
        int[] frequency=new int[6];
        for (int i=0; i<20; i++) {
            frequency[answers[i]]++;
        }
        System.out.println("分值\t学生数");
        for (int i=1; i<6; i++) {
            System.out.println(i+"\t"+frequency[i]);
        }
    }
}
```

3)运行程序

观察结果,如图 5-2 所示。

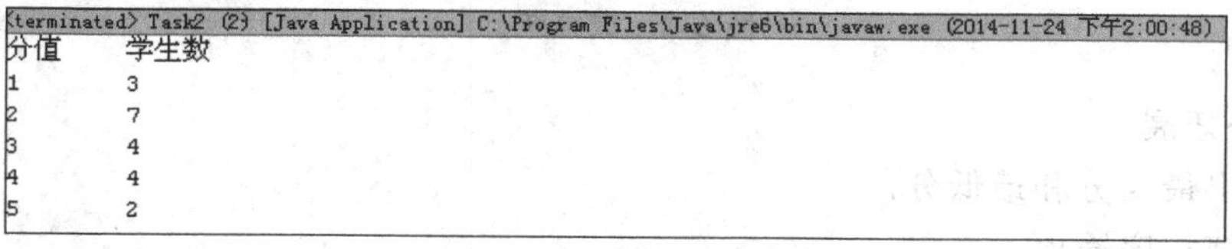

```
<terminated> Task2 (2) [Java Application] C:\Program Files\Java\jre6\bin\javaw.exe (2014-11-24 下午2:00:48)
分值    学生数
1       3
2       7
3       4
4       4
5       2
```

图 5-2 食堂饭菜质量评价

4. 任务拓展

输出对食堂饭菜质量的最终评价分数。

5.3.3 任务 3 打印杨辉三角形

1. 任务目的

（1）理解并掌握二维数组的创建和使用。
（2）能够灵活运用二维数组解决实际问题。

2. 任务描述

杨辉三角形（又称为贾宪三角形，帕斯卡三角形）是二项式系数在三角形中的一种几何排列。要求打印如图 5-3 所示的杨辉三角形。

```
<terminated> Task3 (2) [Java Application] C:\Program Files\Java\jre6\bin\javaw.exe (2014-11-24 下午2:21:55)
1
1 1
1 2 1
1 3 3 1
1 4 6 4 1
1 5 10 10 5 1
```

图 5-3 杨辉三角形

3. 实施步骤

1）算法分析

将图 5-3 可以看成由行和列组成的，用 i 表示行，用 j 表示列，均从 0 开始，分析图 5-3 可以发现如下规律：

如果 j 为 0 时或者对角线上，即 i==j 时数字为 1。
其他情况 a[i][j]=a[i-1][j-1]+a[i-1][j]。

2）参考代码

```java
public class Task3 {
    public static final int ROW=6;                      //设置行数
    public static void main(String[] args) {
        int a[][]=new int[ROW][];
        for (int i=0; i<ROW; i++) {                     //循环初始化数组
            a[i]=new int[i+1];
        }
        for (int i=0; i<ROW; i++) {                     //循环行数
            for (int j=0; j<=a[i].length-1; j++) {      //在行基础上循环列数
                if (j==0 || i==j)
                    a[i][j]=1;                          //将两侧元素设为 1
                else
                    //元素值为其正上方元素与左上角元素之和
                    a[i][j]=a[i-1][j-1]+a[i-1][j];
            }
        }
        for (int i=0; i<ROW; i++) {                     //循环行数
            for (int j=0; j<=a[i].length-1; j++)
```

```
                //在行基础上循环列数
                System.out.print(a[i][j]+" ");         //输出
            System.out.println();                       //换行
        }
    }
}
```

3) 运行程序

观察结果，如图 5-3 所示。

4. 任务拓展

输出等腰三角形形状的杨辉三角形。

第 6 章 类 和 对 象

6.1 实验目的

(1) 理解和掌握面向对象的设计过程。
(2) 会用类图进行面向对象设计。
(3) 掌握类的结构和定义过程。
(4) 掌握对象的创建和使用。
(5) 掌握构造方法及其重载。
(6) 掌握封装的实现及好处。
(7) 掌握 static 关键字、包和访问控制修饰符使用。

6.2 实验任务

(1) 任务 1：手机类的封装。
(2) 任务 2：基于控制台的购书系统。
(3) 任务 3：简单投票程序。

6.3 实验内容

6.3.1 任务 1 手机类的封装

1. 任务目的

(1) 理解和掌握面向对象的设计过程。
(2) 掌握类的结构和定义过程。
(3) 掌握构造方法及其重载。
(4) 掌握对象的创建和使用。

2. 任务描述

参考图 6-1，使用面向对象的思想模拟手机类，编写测试类，使用手机类创建对象，测试手机的各个属性和功能。

3. 实施步骤

1) 任务分析

通过对现实中手机的分析，手机类(Phone)具有以下属性和功能。

(1) 具有属性：品牌(brand)、型号(type)、价格

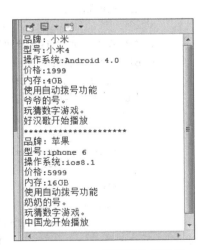

图 6-1 Phone 类测试输出界面

(price)、操作系统(os)和内存(memory)。

(2) 具有功能：查看手机信息(about())、打电话(call(String no))、玩游戏(比如玩猜数字游戏)。

2) UML 类图设计

通过上面的分析,可把手机类使用 UML 类图表示成如图 6-2 所示。

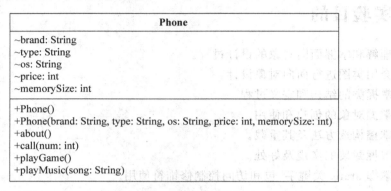

图 6-2　Phone 类 UML 类图

3) 代码实现

手机类 Phone.java 参考代码如下：

```java
package ch7.task1;
public class Phone {
    String brand;                    //品牌
    String type;                     //型号
    String os;                       //操作系统
    int price;                       //价格
    int memorySize;                  //内存
    //无参构造方法
    public Phone() {
    }
    //有参构造方法,进行属性初始化
    public Phone(String brand,String type,String os,int price,int memorySize) {
        this.brand=brand;
        this.type=type;
        this.os=os;
        this.price=price;
        this.memorySize=memorySize;
    }
    //关于本机
    public void about() {
        System.out.println("品牌："+brand+"\n 型号："+type+"\n 操作系统："+os+"\n 价格："+price+"\n 内存："+memorySize+"GB");
    }
    //打电话方法
```

```java
    public void call(int num){
        System.out.println("使用自动拨号功能");
        String phoneNo="";
        switch(num){
        case 1:phoneNo="爷爷的号。";break;
        case 2:phoneNo="奶奶的号。";break;
        case 3:phoneNo="爸爸的号。";break;
        case 4:phoneNo="妈妈的号。";break;
        }
        System.out.println(phoneNo);
    }
    //玩游戏方法
    public void playGame(){
        System.out.println("玩猜数字游戏。");
    }
    //下载音乐方法
    public void downloadMusic(String song){
        System.out.println(song+"开始下载…");
        System.out.println(song+"下载完毕");
    }
    //播放音乐方法
    public void playMusic(String song){
        System.out.println(song+"开始播放");
    }
}
```

测试类 PhoneTest.java 代码参考如下：

```java
package ch6.task1;
public class PhoneTest {
    public static void main(String[] args) {
        //调用无参构造方法创建手机对象 phone1
        Phone phone1=new Phone();
        //phone1 对象通过"."运算符调用自己的属性并赋值
        phone1.brand="小米";
        phone1.type="小米 4";
        phone1.price=1999;
        phone1.os="Android 4.0";
        phone1.memorySize=4;
        //对 phone1 的各项功能进行测试
        phone1.about();
        phone1.call(1);
        phone1.playGame();
        phone1.playMusic("好汉歌");
        System.out.println("*********************");
        //调用有参构造方法创建手机对象 phone2,同时为手机属性赋值
```

```
        Phone phone2=new Phone("苹果","iphone 6","ios8.1",5999,16);
        //对phone2各项功能进行测试
        phone2.about();
        phone2.call(2);
        phone2.playGame();
        phone2.playMusic("中国龙");
    }
}
```

4) 运行程序

观察结果,如图6-1所示。

4. 任务拓展

(1) 手机的playGame()方法中实现真正的猜数字,该如何实现呢?

提示:一种方式,在playGame()方法体内把前面猜数字游戏的实现重新写一遍。另一种方式,把猜数字游戏单独封装为一个类,在playGame()方法体内创建猜数字游戏类对象并调用相应方法实现猜数字。思考一下,哪种方式更好呢?

(2) 可以继续给手机添加其他功能,比如计算器等,同样可以单独封装计算器类,在手机的计算方法中调用。

(3) 通过Phone类的实现过程体会面向对象设计思想。

6.3.2 任务2 基于控制台的购书系统

1. 任务目的

(1) 理解和掌握面向对象的设计过程。

(2) 会用类图进行面向对象设计。

(3) 掌握封装的实现及好处。

(4) 包和访问控制修饰符的使用。

2. 任务描述

开发基于控制台的购书系统。

(1) 输出所有图书的信息:包括每本图书的图书编号、图书名称、图书单价和库存数量。

(2) 顾客购买图书:根据提示输入图书编号来购买图书,并根据提示输入购买图书数量(用户必须连续购书三次)。

(3) 购书完毕后输出顾客的订单信息:包括订单号、订单明细和订单总额。

运行结果参考图6-3。

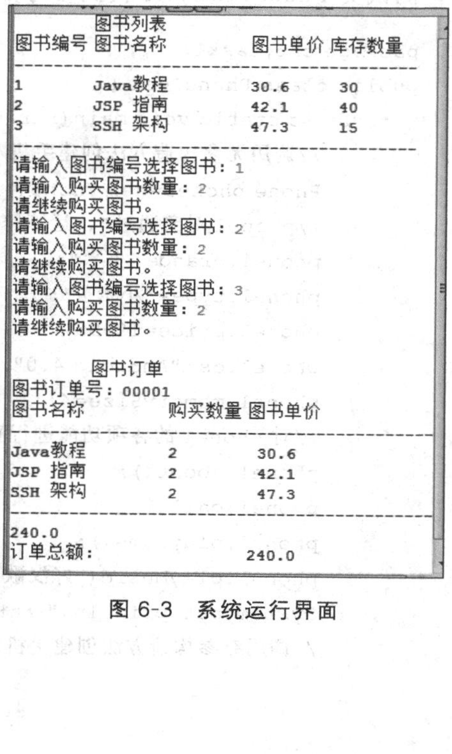

3. 实施步骤

1) 任务分析

该系统中必须包括3个实体类,类名及属性设置如下:

图6-3 系统运行界面

图书类(Book)
    图书编号(id)
    图书名称(name)
    图书单价(price)
    库存数量(storage)
订单项类(OrderItem)
    图书 (book)
    购买数量(num)
订单类(Order):
    订单号(orderId)
    订单总额(total)
    订单项列表(items)

2) UML 类图设计

购书系统 UML 类图如图 6-4 所示。

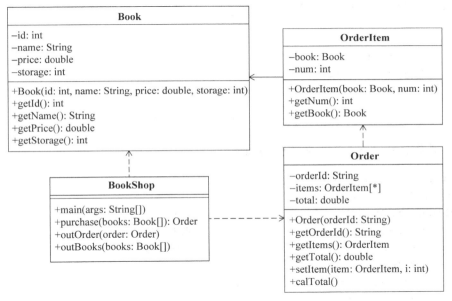

图 6-4　购书系统 UML 类图

3) 代码实现

图书实体类 Book.java 参考代码如下:

```java
package task1.bookshop;
//图书类
public class Book {
    private int id;              //图书编号
    private String name;         //图书名称
    private double price;        //图书单价
    private int storage;         //图书库存
    //有参构造方法
```

```java
    public Book(int id,String name,double price,int storage) {
        this.id=id;
        this.name=name;
        this.price=price;
        this.storage=storage;
    }
    //获取图书编号
    public int getId() {
        return id;
    }
    //获取图书名称
    public String getName() {
        return name;
    }
    //获取图书单价
    public double getPrice() {
        return price;
    }
    //获取图书库存
    public int getStorage() {
        return storage;
    }
}
```

订单项类 OrderItem.java 参考代码如下：

```java
package task1.bookshop;
//订单项类
public class OrderItem {
    private Book book;                    //图书对象
    private int num;                      //订购图书数量
    //有参构造方法
    public OrderItem(Book book,int num) {
        this.book=book;
        this.num=num;
    }
    //获取订购图书数量
    public int getNum() {
        return num;
    }
    //获取图书对象
    public Book getBook() {
        return book;
    }
}
```

订单类 Order.java 参考代码如下：

```java
package task1.bookshop;
//订单类
public class Order {
    private String orderId;                    //订单号
    private OrderItem items[];                 //订单项列表
    private double total;                      //订单总额
    //有参构造方法
    public Order(String orderId) {
        this.orderId=orderId;
        this.items=new OrderItem[3];           //为订单项数组分配空间
    }
    //获取订单号
    public String getOrderId() {
        return orderId;
    }
    //获取订单列表
    public OrderItem[] getItems() {
        return items;
    }
    //获取订单总额
    public double getTotal() {
        calTotal();
        return total;
    }
    //指定一个订单项
    public void setItem(OrderItem item,int i) {
        this.items[i]=item;
    }
    //计算订单总额
    public void calTotal() {
        double total=0;
        for (int i=0; i<items.length; i++) {
            total+=items[i].getNum() * items[i].getBook().getPrice();
        }
        this.total=total;
    }
}
```

主类 BookShop.java 参考代码如下：

```java
package task1.bookshop;
import java.util.Scanner;
//图书商店类
public class BookShop {
    public static void main(String[] args) {
        Book books[]=new Book[3];
```

```java
        //1.模拟从数据库中读出图书信息并输出
        outBooks(books);
        //2.顾客购买图书
        Order order=purchase(books);
        //3.输出订单信息
        outOrder(order);
    }
    //顾客购买图书
    public static Order purchase(Book books[]) {
        Order order=new Order("00001");
        OrderItem item=null;
        Scanner in=new Scanner(System.in);
        for (int i=0; i<3; i++) {
            System.out.print("请输入图书编号选择图书:");
            int cno=in.nextInt();
            System.out.print("请输入购买图书数量:");
            int pnum=in.nextInt();
            item=new OrderItem(books[cno-1],pnum);
            order.setItem(item,i);
            System.out.println("请继续购买图书。");
        }
        return order;
    }
    //输出订单信息
    public static void outOrder(Order order) {
        System.out.println("\n\t图书订单");
        System.out.println("图书订单号:"+order.getOrderId());
        System.out.println("图书名称\t\t购买数量\t图书单价");
        System.out.println("--------------------------------------------");
        OrderItem items[]=order.getItems();
        for (int i=0; i<items.length; i++) {
            System.out.println(items[i].getBook().getName()+"\t"+items[i].getNum()
                +"\t "+items[i].getBook().getPrice());
        }
        System.out.println("--------------------------------------------");
        System.out.println("订单总额:\t\t"+order.getTotal());
    }
    //模拟从数据库中读出图书信息并输出
    public static void outBooks(Book books[]) {
        books[0]=new Book(1,"Java 教程 ",30.6,30);
        books[1]=new Book(2,"JSP 指南 ",42.1,40);
        books[2]=new Book(3,"SSH 架构 ",47.3,15);
        System.out.println("\t图书列表");
        System.out.println("图书编号\t图书名称\t\t图书单价\t库存数量");
```

```
            System.out.println("----------------------------------------");
            for (int i=0; i<books.length; i++) {
                System.out.println(i+1+"\t"+books[i].getName()+"\t"
                        +books[i].getPrice()+"\t "+books[i].getStorage());
            }
            System.out.println("----------------------------------------");
        }
}
```

4) 运行程序

观察结果,如图 6-3 所示。

4. 任务拓展

上例中图书购买次数为 3 次,可不可以修改程序,实现读者按照自己需求决定购买图书次数?

6.3.3 任务 3 简单投票程序

1. 任务目的

(1) 掌握 static 关键字的使用。

(2) 区分实例变量和类变量、实例方法和类方法的区别。

2. 任务描述

编程实现一个投票程序,实现选民投票,每个选民只能投一次票,当投票总数达到 100 时或者主观结束投票时投票程序结束,同时统计投票选民和投票结果。程序运行结果如图 6-5 所示。

图 6-5 程序运行结果

3. 实施步骤

1) 任务分析

从任务描述中抽象出选民 Voter 类,它具有姓名、最大投票数、当前投票总数和投票意见。因为所有选民都会改变同一个数据,即投票次数,因此把它定义为静态变量:

```
private static int count;                    //投票数
```

另外,为了防止选民重复投票,必须保存已经参与投票的选民信息,可采用一个集合来存放已经投票的选民对象。

```
private static Set<Voter>voters=new HashSet<Voter>();      //存放已经投票的选民
```

最后,编写测试 Voter 类的投票和打印投票结果功能。

注:关于 Set 集合的用法可参考第 13 章 Java 集合框架部分,在此作为一个容器来存放选民,而且放入的对象不能重复。

2) 代码实现

Voter.java 类参考代码如下:

```
package task1.statictest;
```

```java
import java.util.HashSet;
import java.util.Set;
public class Voter {
    /**属性的定义*/
    private static final int MAX_COUNT=100;        //静态常量,最大投票数
    private static int count;                       //静态变量,投票数
    //静态变量,存放已经投票的选民
    private static Set<Voter>voters=new HashSet<Voter>();
    private String name;                            //实例变量,投票人
    private String answer;                          //实例变量,投票意见
    /**构造方法*/
    public Voter(String name){
        this.name=name;
    }
    /**投票*/
    public void voterFor(String answer){
        if(count==MAX_COUNT){
            System.out.println("投票结束");
            return;
        }
        if(voters.contains(this))
            System.out.println(name+":你不允许重复投票");
        else {
            this.answer=answer;
            count++;
            voters.add(this);
            System.out.println(name+":感谢你投票");
        }
    }
    /**打印投票结果*/
    public static void printVoterResult(){
        System.out.println("当前投票数为:"+count);
        System.out.println("参与投票的选民和结果如下");
        for(Voter voter:voters){
            System.out.println(voter.name+"意见:"+voter.answer);
        }
    }
    /**main()方法*/
    public static void main(String[] args){
        //创建选民对象
        Voter tom=new Voter("Tom");
        Voter jack=new Voter("Jack");
        Voter mike=new Voter("Mike");
        //选民开始投票
```

```
        tom.voterFor("是");
        tom.voterFor("否");
        jack.voterFor("是");
        mike.voterFor("是");
        Voter.printVoterResult();            //打印投票结果
    }
}
```

3）程序运行结果

运行结果如图 6-5 所示。

第7章 继 承

7.1 实验目的

(1) 掌握继承的实现和继承的作用。
(2) 掌握方法重写。
(3) 掌握继承关系中的构造方法和子类对象的构造过程。
(4) 掌握 this、super 和 final 关键字的使用。
(5) 掌握 toString 方法的使用。

7.2 实验任务

(1) 任务1：公司雇员类封装。
(2) 任务2：汽车租赁系统。
(3) 任务3：饲养员喂养动物。

7.3 实验内容

7.3.1 任务1 公司雇员类封装

1. 任务目的

(1) 掌握继承的实现和继承的作用。
(2) 掌握方法重写。
(3) 掌握继承关系中的构造方法和子类对象的构造过程。
(4) 掌握 this、super 关键字的使用。

2. 任务描述

某公司所有员工根据领取薪金的方式分为如下几类：时薪工(hourlyworker)、管理人员(manager)。时薪工按工作的小时支付工资，每月工作超出160小时的部分按照1.5倍工资发放。管理人员按照级别不同得到固定的工资。编制一个程序来实现该公司的所有员工类，并加以测试。运行结果参考图7-1。

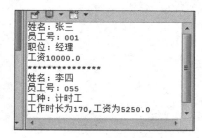

图 7-1 运行结果

3. 实施步骤

1) 任务分析

通过任务描述得知，无论时薪工还是管理人员都属于公司雇员，所以可以定义雇员类(Employee)作为父类，时薪工(HourlyEmployee)和管理人员(Manager)作为子类，相关类具体需求如下。

雇员类(Employee)具有的属性包括姓名(name)、工号(no)和薪水(salary),具有的方法包括构造方法(为 name、no 属性赋值)、打印信息。

时薪工(HourlyEmployee)继承父类的属性和方法的同时,新增 salaryPerHour、hourPerMonth 属性,同时新增计算工资方法、重写打印信息方法。

管理人员(Manager)继承父类的属性和方法的同时,新增 level 属性,同时新增计算工资方法、重写打印信息方法。

2) UML 类图设计

根据以上分析,使用 UML 类图设计如图 7-2 所示。

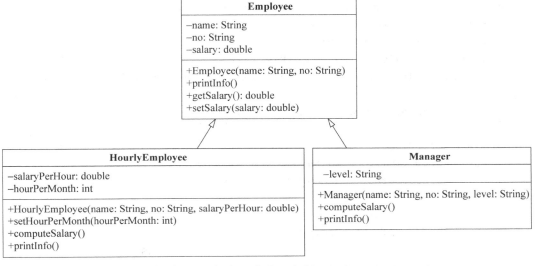

图 7-2 雇员类继承关系 UML 图

3) 代码实现

父类 Employee.java 参考代码如下:

```java
package ch7.task1;
public class Employee {
    private String name;                    //雇员姓名
    private String no;                      //雇员工号
    private double salary;                  //雇员薪水
    //构造方法,初始化姓名、工号
    public Employee(String name,String no) {
        this.name=name;
        this.no=no;
    }
    //打印信息方法
    public void printInfo(){
        System.out.println("姓名: "+name+"\n员工号: "+no);
    }
```

```java
    //属性 salary 的 getter/setter 方法
    public double getSalary() {
        return salary;
    }
    public void setSalary(double salary) {
        this.salary=salary;
    }
}
```

子类 HourlyEmployee.java 参考代码如下:

```java
package ch7.task1;
public class HourlyEmployee extends Employee {
    private double salaryPerHour;                    //每小时薪水
    private int hourPerMonth;                        //月工作时数
    //构造方法,初始化姓名、工号、每小时薪水
    public HourlyEmployee(String name,String no,double salaryPerHour) {
        super(name,no);
        this.salaryPerHour=salaryPerHour;
    }
    //设定月工作时数
    public void setHourPerMonth(int hourPerMonth) {
        this.hourPerMonth=hourPerMonth;
    }
    //计时工计算工资方法
    public void computeSalary() {
        if (this.hourPerMonth<160) {                 //160 时之内
            setSalary(salaryPerHour * this.hourPerMonth);
        } else {                                     //超过 160 小时
            setSalary((this.salaryPerHour * (hourPerMonth-160) * 1.5)+
                (160 * this.salaryPerHour));
        }
        System.out.println("工作时长为"+hourPerMonth+",工资为"+getSalary());
    }
    //重写打印信息方法
    public void printInfo(){
        super.printInfo();
        System.out.println("工种:计时工");
    }
}
```

子类 Manager.java 参考代码如下:

```java
package ch7.task1;
public class Manager extends Employee{
    private String level;                            //管理者级别
    //构造方法,初始化姓名、工号、级别
```

```java
    public Manager(String name,String no,String level) {
        super(name,no);
        this.level=level;
    }
    //管理者工资计算方法
    public void computeSalary(){
        if(level.equals("经理")){
            setSalary(10000);
        }else if(level.equals("副经理")){
            setSalary(6000);
        }else if(level.equals("车间主任")){
            setSalary(4000);
        }else{
            setSalary(3000);
        }
        System.out.println("工资"+getSalary());
    }
    //重写打印信息方法
    public void printInfo(){
        super.printInfo();
        System.out.println("职位: "+level);
    }
}
```

测试类 Test.java 参考代码如下：

```java
package ch7.task1;
public class Test {
    public static void main(String[] args) {
        Manager m=new Manager("张三","001","经理");
        m.printInfo();
        m.computeSalary();
        System.out.println("***************");
        HourlyEmployee he=new HourlyEmployee("李四","055",30);
        he.printInfo();
        he.setHourPerMonth(170);
        he.computeSalary();
    }
}
```

4）运行程序

观察结果，如图 7-1 所示。

4. 任务拓展

继续添加计件工、销售人员等雇员类，体会继承带来的好处。

7.3.2 任务2 汽车租赁系统

1. 任务目的

（1）掌握继承的实现和继承的作用。

（2）掌握方法重写。

（3）掌握继承关系中的构造方法和子类对象的构造过程。

（4）掌握 this、super 关键字的使用。

2. 任务描述

某汽车租赁公司出租多种轿车和客车，出租费以日为单位计算，不同车型日租金情况如表 7-1 所示。

表 7-1 不同车型日租金情况

类 别	轿 车			客 车	
车型	商务舱 GL8	宝马 550i	林荫大道	≤19 座	>19 座
日租费（元/天）	600	500	300	800	1200

采用面向对象的思想编程实现计算不同车型不同天数的租赁费用，运行结果如图 7-3 所示。

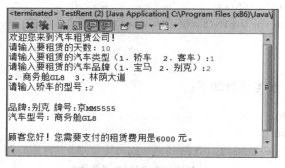

图 7-3 汽车租赁运行结果

3. 实施步骤

1）任务分析

通过任务描述，租赁公司有轿车类和客车类，而它们都具有共同的特征，所以可以抽象出汽车（MotoVehicle）类作为父类，轿车类（Car）和客车（Bus）作为子类。它们主要属性和方法如下。

MotoVehicle 类具有的属性：车牌号（no）、品牌（brand）。

具有的方法：打印汽车信息（printInfo()）、计算租金方法（int calRent(int days)）。

Car 类继承 MotoVehicle 类的属性，同时新增属性：汽车型号（type）。

具有的方法：重写计算租金方法（按汽车型号计算）、重写打印汽车信息方法。

Bus 类继承 MotoVehicle 类的属性，同时新增属性：座位数（seatCount）。

具有的方法：重写计算租金方法（按汽车座位数计算）、重写打印汽车信息方法。

TestRent 类为主类，在其 main() 方法中对汽车租赁系统进行测试。

2）UML 类图设计

汽车租赁 UML 类图如图 7-4 所示。

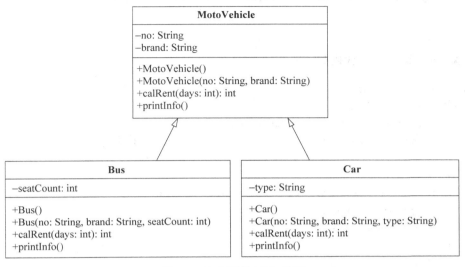

图 7-4 汽车租赁 UML 类图

3）代码实现

父类 MotoVehicle.java 参考代码如下：

```java
package ch7.task2;
/***父类汽车类*/
public class MotoVehicle {
    private String no;                          //汽车牌号
    private String brand;                       //汽车品牌
    /***无参构造方法 */
    public MotoVehicle() {
    }
    /***初始化牌号、品牌有参构造方法*/
    public MotoVehicle(String no,String brand) {
        this.no=no;
        this.brand=brand;
    }
    public String getNo() {
        return no;
    }
    public String getBrand() {
        return brand;
    }
    /***计算汽车租赁价 */
    public int calRent(int days){
        return 0;
    }
```

```java
/**打印汽车品牌和车牌号*/
public void printInfo(){
    System.out.println("\n 品牌:"+brand+" 牌号:"+no);
}
}
```

子类 Car.java 参考代码如下：

```java
package ch7.task2;
/**轿车类,继承汽车类*/
public class Car extends MotoVehicle {
    private String type;                              //汽车型号
    public Car() {
    }
    public Car(String no,String brand,String type) {
        super(no,brand);
        this.type=type;
    }
    public String getType() {
        return type;
    }
    public void setType(String type) {
        this.type=type;
    }
    /**重写父类计算客车租赁价方法*/
    public int calRent(int days) {
        if (type.equals("550i")) {                    //车型是宝马 550i
            return days * 500;
        } else if (type.equals("商务舱 GL8")) {      //车型是别克商务舱 GL8
            return 600 * days;
        } else {
            return 300 * days;
        }
    }
    /**重写父类打印汽车信息方法*/
    public void printInfo(){
        super.printInfo();
        System.out.println("汽车型号："+type);
    }
}
```

子类 Bus.java 参考代码如下：

```java
package ch7.task2;
/**客车类,继承汽车类*/
public final class Bus extends MotoVehicle {
    private int seatCount;                            //座位数
```

```java
    public Bus() {
    }
    public Bus(String no,String brand,int seatCount) {
        super(no,brand);
        this.seatCount=seatCount;
    }
    public int getSeatCount() {
        return seatCount;
    }
    public void setSeatCount(int seatCount) {
        this.seatCount=seatCount;
    }
    /**重写父类计算客车租赁价方法*/
    public int calRent(int days) {
        if (seatCount<=19) {
            return days * 800;
        } else {
            return days * 1500;
        }
    }
    /**重写父类打印汽车信息方法*/
    public void printInfo(){
        super.printInfo();
        System.out.println("客车座位数:"+seatCount);
    }
}
```

测试类 TestRent.java 参考代码如下：

```java
package ch7.task2;
import java.util.Scanner;
/**主类 TestRent*/
public class TestRent {
    public static void main(String[] args) {
        String no,brand,answer,type;
        int seatCount,days,rent;
        Car car;
        Bus bus;
        Scanner input=new Scanner(System.in);
        System.out.println("欢迎您来到汽车租赁公司!");
        System.out.print("请输入要租赁的天数:");
        days=input.nextInt();
        System.out.print("请输入要租赁的汽车类型(1.轿车  2.客车):");
        answer=input.next();
        if("1".equals(answer)){
            System.out.print("请输入要租赁的汽车品牌(1.宝马  2.别克):");
            answer=input.next();
```

```java
            if("1".equals(answer)){
                System.out.println("1. 550i: ");
                brand="宝马";}
            else{
                brand="别克";
                System.out.print("2. 商务舱 GL8   3. 林荫大道");
            }
            System.out.println("请输入轿车的型号:");
            answer=input.next();
            if("1".equals(answer)){
                type="550i";
            }else if("2".equals(answer)){
                type="商务舱 GL8";
            }else{
                type="林荫大道";
            }
            no="京 MM5555";                    //简单起见,直接指定汽车牌号
            car=new Car(no,brand,type);
            car.printInfo();
            rent=car.calRent(days);
        }
        else{
            System.out.print("请输入要租赁的客车品牌(1. 金杯   2. 金龙):");
            answer=input.next();
            if("1".equals(answer)){
                brand="金杯";
            }else{
                brand="金龙";
            }
            System.out.print("请输入客车的座位数:");
            seatCount=input.nextInt();
            no="京 GG8888";
            bus=new Bus(no,brand,seatCount);
            bus.printInfo();
            rent=bus.calRent(days);
        }
    System.out.println("\n 顾客您好!您需要支付的租赁费用是"+rent+"元。");
    }
}
```

4)运行程序

观察结果,如图 7-3 所示。

4. 任务拓展

(1) 继续添加卡车类,可以根据承载的吨位收取租赁费用。

品牌有福田、江淮、东风等,租赁费按照 3 吨以上每天 1500 元,3 吨以下每天 1000 元等。

(2) 可以考虑添加顾客类,可以进行汽车租赁。

7.3.3 任务3 饲养员喂养动物

1. 任务目的

(1) 掌握继承的实现和继承的作用。
(2) 掌握继承关系中的构造方法和子类对象的构造过程。
(3) 掌握方法重写。
(4) 掌握 super、this、toString 方法的使用。

2. 任务描述

动物园中饲养员可以拿着不同的食物喂养不同的动物,编程实现饲养员喂养动物程序,要求如下。

(1) 饲养员(Feeder)可以给动物(Animal)喂食物(Food)。
(2) 现在动物有 Dog 和 Cat,食物有 Fish 和 Bone。
(3) 饲养员可以给狗喂骨头,给猫喂鱼。

运行结果参考图 7-5。

图 7-5 程序运行结果

3. 实施步骤

1) 任务分析

为了实现以上需求,需要定义以下 8 个类,类说明如下。

(1) 动物类父类 Animal 类具有 eat()方法,代表各种动物吃的共性。
(2) Animal 子类 Dog 继承父类的 eat()方法,新增 eat(Bone bone)方法,同时重写 toString()方法。
(3) Animal 子类 Cat 继承父类的 eat()方法,新增 eat(Fish fish)方法,同时重写 toString()方法。
(4) 食物类的父类 Food,具有 weight 属性。
(5) Food 子类 Bone 继承父类的 weight 方法,调用父类构造方法初始化 weight 属性,同时重写 toString 方法。
(6) Food 子类 Fish 继承父类的 weight 方法,调用父类构造方法初始化 weight 属性,同时重写 toString()方法。
(7) 饲养员类 Feeder 具有 name 属性,同时具有重载方法 feed(Dog dog,Bone bone)和 feed(Cat cat,Fish fish)。
(8) 测试类 Test 测试 Feeder 的喂养方法。

2) UML 类图设计

饲养员喂养动物 UML 的类图如图 7-6 所示。

3) 代码实现

父类 Animal.java 参考代码如下:

```
package ch7.task3;
/**动物父类 Animal*/
```

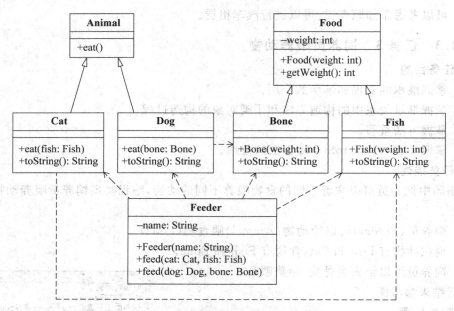

图 7-6 饲养员喂养动物 UML 的类图

```
public class Animal {
    public void eat(){
        System.out.println("吃饭时间到了,");
    }
}
```

Animal 子类 Dog.java 参考代码如下：

```
package ch7.task3;
/**Animal 子类 Dog*/
public class Dog extends Animal {
    //子类新增 eat()方法,因为跟父类 eat()方法参数不同,所以不属于重写
    public void eat(Bone bone){
        eat();                  //调用从父类继承的 eat()方法
        System.out.println(this+"喜欢吃"+bone);
    }
    //重写 Animal 从 Object 中继承的 toString()方法
    public String toString(){
        return "Dog";
    }
}
```

Animal 子类 Cat.java 参考代码如下：

```
package ch7.task3;
/**Animal 子类 Cat*/
public class Cat extends Animal {
    //子类新增 eat()方法,因为跟父类 eat()方法参数不同,所以不属于重写
```

```java
    public void eat(Fish fish){
        eat();                     //调用从父类继承的eat()方法
        System.out.println(this+"喜欢吃"+fish);
    }
    //重写Animal从Object中继承的toString()方法
    public String toString(){
        return "Cat";
    }
}
```

父类Food.java参考代码如下：

```java
package ch7.task3;
/**食物父类Food*/
public class Food {
    private int weight;          //食物重量
    public Food(int weight) {
        this.weight=weight;
    }
    public int getWeight(){
        return weight;
    }
}
```

Food子类Bone.java参考代码如下：

```java
package ch7.task3;
/**Food子类Bone*/
public class Bone extends Food {
    public Bone(int weight) {
        super(weight);           //调用父类的构造方法
    }
    //重写Food从Object中继承的toString()方法
    public String toString(){
        return "Bone";
    }
}
```

Food子类Fish.java参考代码如下：

```java
package ch7.task3;
/**Food子类Fish*/
public class Fish extends Food {
    public Fish(int weight) {
        super(weight);           //调用父类构造方法
    }
    //重写Food从Object中继承的toString()方法
    public String toString(){
```

```java
        return "Fish";
    }
}
```

饲养员类 Feeder.java 参考代码如下:

```java
package ch7.task3;
/**饲养员类 Feeder*/
public class Feeder {
    private String name;           //饲养员姓名
    public Feeder(String name){
        this.name=name;
    }
    //拿鱼喂养猫方法
    public void feed(Cat cat,Fish fish){
        cat.eat(fish);
        System.out.println("饲养员"+name+"拿着"+fish.getWeight()+"克"+fish+
                "喂养"+cat+"。");
    }
    //拿骨头喂养狗方法
    public void feed(Dog dog,Bone bone){
        dog.eat(bone);
        System.out.println("饲养员"+name+"拿着"+bone.getWeight()+"克"+bone+
                "喂养"+dog+"。");
    }
}
```

测试类 TestFeed.java 参考代码如下:

```java
public class TestFeed {
    public static void main(String[] args) {
        Feeder feeder=new Feeder("张三");
        Dog dog=new Dog();
        Bone bone=new Bone(500);
        feeder.feed(dog,bone);                         //喂养 Dog
        feeder.feed(new Cat(),new Fish(300));          //喂养 Cat
    }
}
```

4) 运行程序

观察结果,如图 7-5 所示。

4. 任务拓展

(1) 继续添加老虎类(Tiger)和肉类(Meat),分别继承 Animal 类和 Food 类,在 Feeder 类中添加喂养方法,实现拿肉喂养老虎。

(2) 体会该程序的优势和弊端。

第8章 多 态

8.1 实验目的

(1) 掌握多态的含义及应用场合。
(2) 掌握上转型对象和多态的实现。
(3) 掌握 instanceof 运算符的使用。
(4) 掌握 abstract 关键字的使用。

8.2 实验任务

(1) 任务1：图形面积周长计算小程序。
(2) 任务2：饲养员喂养动物程序优化。

8.3 实验内容

8.3.1 任务1 图形面积周长计算小程序

1．任务目的

(1) 掌握多态的含义及应用场合。
(2) 掌握上转型对象和多态的实现。
(3) 掌握 abstract 关键字的使用。

2．任务描述

设计一个小程序，可以计算圆形和长方形的面积及周长，其中定义抽象类图形类为圆形和长方形的父类，在图形类中定义抽象方法获取面积方法和获取周长方法。定义面积和周长计算器，可以计算不同图形的面积和周长。程序要具备良好的可扩展性与可维护性。程序运行结果参考图8-1。

图 8-1 程序运行结果

3．实施步骤

1) 任务分析

定义父类 Shape 作为抽象类，并在类中定义抽象方法求周长和求面积。

定义 Shape 子类圆形(Circle)，具有属性半径(radius)和常量 PI，同时必须实现父类中的抽象方法。

定义 Shape 子类长方形(Rectangle)，具有属性长(length)和宽(width)，同时也必须实现父类的抽象方法。

创建图形面积周长计算器(ShapeCaculate),具有计算不同图形面积和周长的方法。创建测试类 TestShape 类,在其 main()方法中对 ShapeCaculate 计算面积和周长方法进行测试。

2) UML 类图设计

图形面积和周长计算 UML 类图如图 8-2 所示。

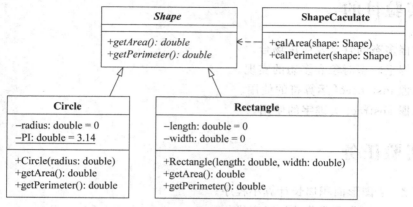

图 8-2　图形面积和周长计算 UML 类图

3) 代码实现

抽象类 Shape.java 参考代码如下:

```java
package task1;
/** 抽象类：几何图形。 */
abstract class Shape {
    //抽象方法：获取面积
    public abstract double getArea();
    //抽象方法：获取周长
    public abstract double getPerimeter();
}
```

子类 Circle.java 参考代码如下:

```java
package task1;
/**圆形 */
class Circle extends Shape {
    private double radius=0;                         //圆的半径
    private final static double PI=3.14;      //常量,圆周率
    //有参构造方法,初始化圆半径
    public Circle(double radius) {
        this.radius=radius;
    }
    //求圆的面积
    public double getArea() {
        return (PI * radius * radius);
    }
```

```java
        //求圆的周长
        public double getPerimeter() {
            return 2 * PI * radius;
        }
}
```

子类 Rectangle.java 参考代码如下：

```java
package task1;
/**长方形 */
class Rectangle extends Shape {
    private double length=0;                    //长方形的长
    private double width=0;                     //长方形的宽
    //有参构造方法,初始化长方形的长和宽
    public Rectangle(double length,double width) {
        super();
        this.length=length;
        this.width=width;
    }
    //重写父类求面积的方法
    public double getArea() {
        return (this.length * this.width);
    }
    //重写父类求周长的方法
    public double getPerimeter() {
        return 2 * (length+width);
    }
}
```

类 ShapeCaculate.java 参考代码如下：

```java
package task1;
/**图形面积和周长计算器 */
public class ShapeCaculate {
    //可以计算任何 Shape 子类的面积
    public void calArea(Shape shape){
        System.out.println(shape.getArea());
    }
    //可以计算任何 Shape 子类的周长
    public void calPerimeter(Shape shape){
        System.out.println(shape.getPerimeter());
    }
}
```

类 TestShapeCaculate.java 参考代码如下：

```java
package task1;
/** 测试类 */
```

```java
class TestShape {
    public static void main(String[] args) {
        //创建图形计算器
        ShapeCaculate sc=new ShapeCaculate();
        //创建长方形和圆形对象
        Shape rectangle=new Rectangle(3,4);
        Circle circle=new Circle(3);
        //求长方形和圆形面积
        System.out.println("求长方形面积：");
        sc.calArea(rectangle);
        System.out.println("求圆形面积：");
        sc.calArea(circle);
        //求长方形和圆形周长
        System.out.println("求长方形周长：");
        sc.calPerimeter(rectangle);
        System.out.println("求圆形周长：");
        sc.calPerimeter(circle);
    }
}
```

4) 运行程序

观察结果，如图 8-1 所示。

4. 任务拓展

要新增一种图形计算其面积和周长时，不需要修改 ShapeCaculate 类，程序具有良好的可扩展性和可维护性。继续添加三角形、梯形等形状，分别计算面积和周长，体会多态带来的好处。

8.3.2 任务 2 饲养员喂养动物程序优化

1. 任务目的

（1）掌握多态的含义及应用场合。

（2）掌握上转型对象和多态的实现。

（3）体会多态带来的好处。

（4）掌握 instanceof 运算符的使用。

2. 任务描述

对第 7 章的任务 3——饲养员喂养动物进行优化，使程序具备良好的可扩展性和可维护性。

3. 实施步骤

1）任务分析

在第 7 章的任务 3 中，饲养员每拿一种食物喂养相应动物都需要建立相应的方法，程序的可扩展性和可维护性较差，通过多态可以对程序进行优化，修改 feed() 方法的参数为父类的类型，使程序具有较好的可扩展性和可维护性。

2）UML 类图

对其 UML 类图进行优化后的 UML 类图如图 8-3 所示。

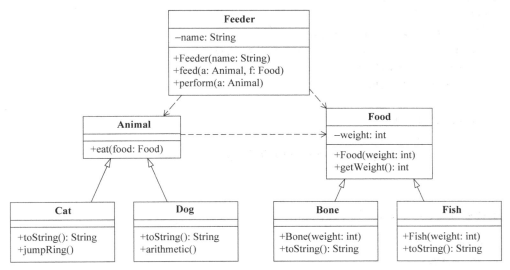

图 8-3 饲养员喂养动物优化后 UML 类图

3）代码实现

父类 Animal.java 参考代码如下：

```
package task3;
/**动物父类 Animal*/
public class Animal {
    public void eat(Food food){
        System.out.print("吃饭时间到了,");
        System.out.println(this+"喜欢吃"+food);
    }
}
```

Animal 子类 Dog.java 参考代码如下：

```
package ch7.task3;
/**Animal 子类 Dog*/
public class Dog extends Animal {
    //重写 Animal 从 Object 中继承的 toString()方法
    public String toString(){
        return "Dog";
    }
    //新增算算术方法
    public void arithmetic(){
        System.out.println(this+"算算术表演!");
    }
}
```

Animal 子类 Cat.java 参考代码如下：

```java
package ch7.task3;
/**Animal 子类 Cat*/
public class Cat extends Animal {
    //重写 Animal 从 Object 中继承的 toString()方法
    public String toString(){
        return "Cat";
    }
    //新增跳环方法
    public void jumpRing(){
        System.out.println(this+"开始表演跳环!");
    }
}
```

饲养员类 Feeder.java 参考代码如下：

```java
package task3;
/**饲养员类 Feeder*/
public class Feeder {
    private String name;                          //饲养员姓名
    public Feeder(String name){
        this.name=name;
    }
    //可以拿不同食物喂养不同动物的方法
    public void feed(Animal a,Food f){
        a.eat(f);
        System.out.println("饲养员"+name+"拿着"+f.getWeight()+"克"+f+"喂养"+a+"。");
    }
    //饲养员训练动物表演
    public void perform(Animal a){
        if(a instanceof Dog){
            ((Dog)a).arithmetic();
        }
        if(a instanceof Cat){
            Cat c=(Cat)a;                         //向下转型
            c.jumpRing();
        }
    }
}
```

测试类 TestFeed.java 参考代码如下：

```java
package task3;
public class TestFeed {
    public static void main(String[] args) {
        Feeder feeder=new Feeder("张三");
```

```
        Dog dog=new Dog();
        Bone bone=new Bone(500);
        //喂养 Dog
        feeder.feed(dog,bone);
        //喂养 Cat
        feeder.feed(new Cat(),new Fish(300));
        //狗狗表演算算术
        feeder.perform(dog);
        //猫咪表演跳环
        feeder.perform(new Cat());
    }
}
```

父类 Food.java、子类 Bone.java、子类 Fish.java 代码同第 7 章任务 3 中的代码，在此略过。

4) 运行程序，观察结果

运行程序，验证任务描述中的结果。

4. 任务拓展

要实现拿肉(Meat)喂养老虎(Tiger)，程序该如何修改？

第 9 章 接 口

9.1 实验目的

(1) 理解并掌握如何定义接口。
(2) 掌握接口的实现方式。
(3) 理解接口与抽象类的区别。
(4) 掌握使用接口回调来简化程序。

9.2 实验任务

(1) 任务 1：设计实现发声接口。
(2) 任务 2：动物乐园。

9.3 实验内容

9.3.1 任务 1 设计实现发声接口

1. 任务目的

(1) 理解并掌握如何定义接口。
(2) 掌握接口的实现方式。

2. 任务描述

设计和实现一个 Soundable 接口，该接口具有发声功能，同时还能调节声音大小。Soundable 接口的这些功能将会由 3 种声音设备来实现，它们分别是 Radio、Walkman 和 Mobilephone。最后还需设计一个应用程序类来使用这些实现了 Soundable 接口的声音设备。程序运行时，先询问用户想听哪种设备，然后程序按照该设备的工作方式来输出发音。

3. 实施步骤

1) 定义 Soundable 接口

参考代码如下：

```
/**
 * 发声接口
 */
public interface Soundable {
    //发出声音
    public void playSound();
    //降低音量
```

```
    public void decreaseVolume();
    //停止发声
    public void stopSound();
}
```

2) 设计 Radio 类

参考代码如下：

```
/**
 * 收音机类
 */
public class Radio implements Soundable{
    public void playSound(){
        System.out.println("收音机播放广播：中央人民广播电台");
    }
    public void decreaseVolume(){
        System.out.println("降低收音机音量");
    }
    public void stopSound(){
        System.out.println("关闭收音机");
    }
}
```

3) 设计 Walkman 类

参考代码如下：

```
/**
 * 随身听
 */
public class Walkman implements Soundable{
    public void playSound(){
        System.out.println("随身听发出音乐");
    }
    public void decreaseVolume(){
        System.out.println("降低随身听音量");
    }
    public void stopSound(){
        System.out.println("关闭随身听");
    }
}
```

4) 设计 MobilePhone 类

参考代码如下：

```
/**
 * 手机
```

```
 */
class Mobilephone implements Soundable{
    public void playSound(){
        System.out.println("手机发出来电铃声：叮当、叮当");
    }
    public void decreaseVolume(){
        System.out.println("降低手机音量");
    }
    public void stopSound(){
        System.out.println("关闭手机");
    }
}
```

5）设计 SampleDisplay 类

参考代码如下：

```
/**
 * 声音采样
 */
class SampleDisplay{
    public void display(Soundable soundable){
        soundable.playSound();
        soundable.decreaseVolume();
        soundable.stopSound();
    }
}
```

6）设计测试驱动类

参考代码如下：

```
public class Task1 {
    public static void main(String[] args) {
        Scanner scanner=new Scanner(System.in);
        System.out.println("你想听什么?请输入选择!");
        System.out.println("0-收音机  1-随身听  2-手机");
        int choice;
        choice=scanner.nextInt();
        SampleDisplay SampleDisplay=new SampleDisplay();
        //将实现该接口的类的实例的引用传递给该接口参数
        if(choice==0)
            SampleDisplay.display(new Radio());
        else if(choice==1)
            SampleDisplay.display(new Walkman());
        else if (choice==2)
            SampleDisplay.display(new Mobilephone());
        else
```

```
        System.out.println("孩子,你输错了!");
    }
}
```

7)运行程序

观察结果,如图9-1所示。

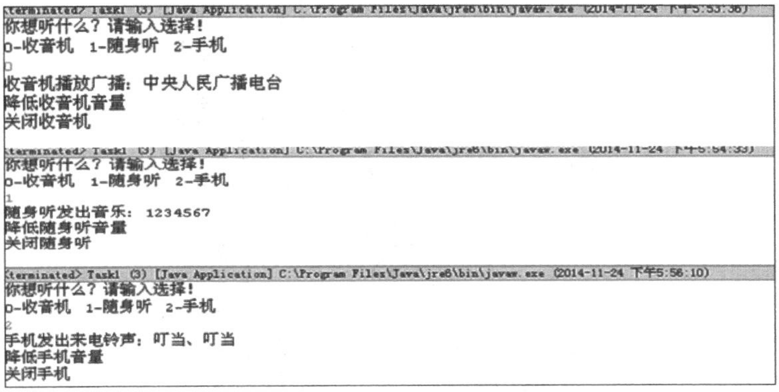

图9-1 设计实现发声接口

4. 任务拓展

(1)思考以上代码哪里发生了接口回调。

(2)增加新的发声设备,比如扩音器。

9.3.2 任务2 动物乐园

1. 任务目的

能够灵活运用接口解决多继承问题。

2. 任务描述

编写程序模拟动物园里饲养员给各种动物喂养各种不同食物的过程,当饲养员给动物喂食时,动物会发出欢快的叫声。

3. 实施步骤

1)问题分析

在这个动物园里,涉及的对象有饲养员,各种不同的动物以及各种不同的食物。这样很容易抽象出3个类Feeder、Animal和Food。假设只考虑猫类和狗类动物,则由Animal类派生出Cat类、Dog类,同样由Food类可以进一步派生出其子类Bone、Fish。因为它们之间存在着明显的is-a关系。

同时鱼是一种食物,但实际上,鱼也是一种动物,Cat类和Dog类继承了Animal的很多属性和方法,如果将Animal定义为接口,Animal中是不能定义成员变量和成员方法的,Food类中虽然也有变量但是数量比Animal少,所以我们考虑将Food定义为接口,此时可以说"鱼是一种动物,同时还可以当作是一种食物"。

2)根据分析,画出UML类图

参考类图如图9-2所示。

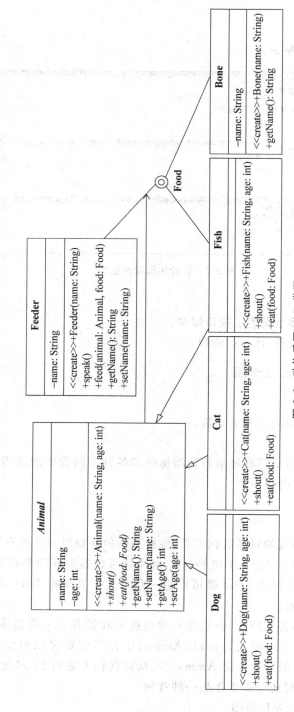

图 9-2 动物乐园 UML 类图

3) 根据类图，定义 Animal 类

参考代码如下：

```java
/**
 * 动物类
 * @author cabbage
 */
public abstract class Animal {
    private String name;
    private int age;

    public Animal(String name,int age) {
        super();
        this.name=name;
        this.age=age;
    }

    public abstract void shout();
    public abstract void eat(Food food);
    public String getName() {
        return name;
    }
    public void setName(String name) {
        this.name=name;
    }
    public int getAge() {
        return age;
    }
    public void setAge(int age) {
        this.age=age;
    }
}
```

4) 定义 Animal 的子类 Cat 类

参考代码如下：

```java
public class Cat extends Animal{
    public Cat(String name,int age) {
        super(name,age);
    }
    public void shout() {
        System.out.println("喵呜…");
    }
    public void eat(Food food) {
        System.out.println(getName()+"正在吃着香喷喷的"+food.getName());
        shout();
    }
}
```

5）定义 Animal 的子类 Dog 类

参考代码如下：

```java
public class Dog extends Animal {
    public Dog(String name,int age) {
        super(name,age);
    }
    public void shout() {
        System.out.println("汪汪汪…");
    }
    public void eat(Food food) {
        System.out.println(getName()+"正在啃着香喷喷的"+food.getName());
        shout();
    }
}
```

6）定义 Food 接口

参考代码如下：

```java
/**
 * 食物接口
 * @author cabbage
 */
public interface Food {
    public abstract String getName();
}
```

7）定义 Fish 类

参考代码如下：

```java
//鱼既是一种动物,同时还可以作为食物
public class Fish extends Animal implements Food {
    public Fish(String name,int age) {
        super(name,age);
    }
    public void shout() {
    }
    public void eat(Food food) {
    }
}
```

8）定义 Bone 类

因为 Bone 类与 Fish 类非常相似,所以此处没有给出参考代码。

9）定义 Feeder 类

参考代码如下：

```java
public class Feeder {
```

```java
    private String name;
    public Feeder(String name) {
        super();
        this.name=name;
    }
    public void speak(){
        System.out.println("欢迎来到动物园");
        System.out.println("我是饲养员"+getName());
    }
    public void feed(Animal animal,Food food){
        animal.eat(food);
    }
    public String getName() {
        return name;
    }
    public void setName(String name) {
        this.name=name;
    }
}
```

10）定义测试驱动类

```java
public class ZooDemo {
    public static void main(String[] args) {
        Feeder feeder=new Feeder("华华");
        feeder.speak();
        Dog dog=new Dog("乐乐",3);
        Food food=new Bone("骨头");
        feeder.feed(dog,food);
        Cat cat=new Cat("甜甜",2);
        food=new Fish("黄花鱼",0);
        feeder.feed(cat,food);
    }
}
```

11）运行程序

观察结果，如图 9-3 所示。

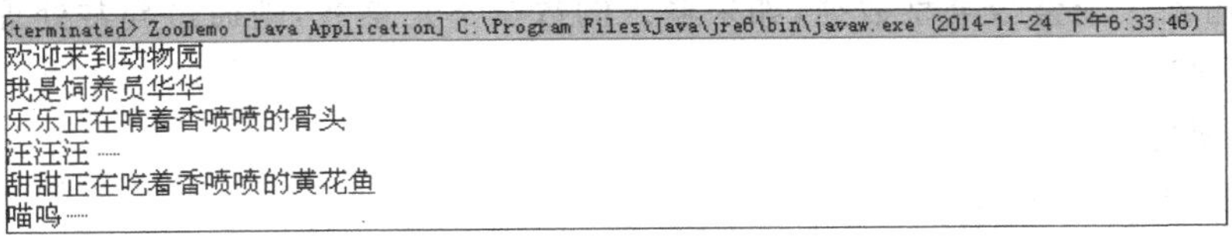

```
<terminated> ZooDemo [Java Application] C:\Program Files\Java\jre6\bin\javaw.exe (2014-11-24 下午6:33:46)
欢迎来到动物园
我是饲养员华华
乐乐正在啃着香喷喷的骨头
汪汪汪
甜甜正在吃着香喷喷的黄花鱼
喵呜
```

图 9-3 动物乐园模拟

4. 任务拓展

在动物园里再加入更多种动物。

第 10 章　异 常 处 理

10.1　实验目的

（1）掌握使用 try-catch 结构进行异常处理的方法。
（2）掌握 finally 代码块的用法。
（3）掌握 throws 与 throw 的用法及区别。
（4）掌握自定义异常的定义与使用。
（5）掌握 JDK 1.7 中的 try-with-resource 结构的用法。
（6）掌握 JDK 1.7 中的 multi-catch 结构及 RethrowException 结构的用法。

10.2　实验任务

（1）任务 1：判断从键盘输入的整数是否合法。
（2）任务 2：处理除数为 0 的异常。
（3）任务 3：处理数组的下标越界异常。
（4）任务 4：特殊字符检查器。
（5）任务 5：使用 try-with-resource 进行读取文件处理。

10.3　实验内容

10.3.1　任务 1　判断从键盘输入的整数是否合法

1. 任务目的
（1）掌握 try-catch 结构异常处理方法。
（2）掌握输入数据类型不匹配异常 InputMismatchException 的处理方法。
（3）掌握异常处理父类 Exception 的使用方法。

2. 任务描述
从键盘输入一个整数类型数据，如果输入的数据不是整数，则通过异常处理给出提示。

3. 实施步骤
1）创建项目
创建 Java 项目 Lab10。
2）创建包
在项目 Lab10 中创建包 task1。
3）创建文件并进行编辑
在包 task1 中创建 Java 类文件 chkInputInt.java，修改该文件的内容如下：

```java
package task1;
import java.util.InputMismatchException;
import java.util.Scanner;
/**
 * 判断从键盘输入的整数是否合法
 * @author sf */
public class ChkInputInt {
    public static void main(String[] args) {
        Scanner input=new Scanner(System.in);          //输入设备
        int num;                                        //要接收的整数
        System.out.print("请输入一个整数：");
        try{
            num=input.nextInt();
            System.out.println("您所输入的整数为："+num);
        }catch(InputMismatchException ex){
            System.out.println("请输入整数数据。");
        }catch(Exception ex){
            System.out.println("其他类型的异常。");
        }
        input.close();                                  //关闭输入
    }
}
```

4) 运行程序

程序的两个运行实例如图 10-1 所示。

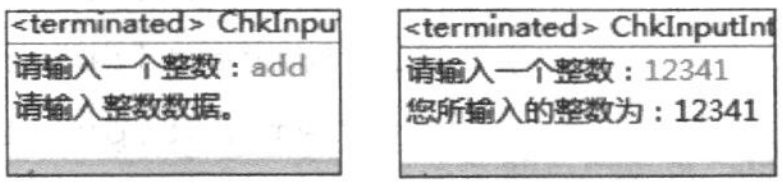

4. 任务扩展

(1) 判断从键盘输入的长整型数据是否合法。

(2) 判断从键盘输入的浮点型、双精度型、字节型等数据是否合法。

图 10-1 判断从键盘输入的整数是否合法运行实例

10.3.2 任务2 处理除数为 0 的异常

1. 任务目的

(1) 掌握 try-catch 结构的异常处理方法。

(2) 掌握数据类型不匹配异常的处理方法。

(3) 掌握除数为 0 的异常处理方法。

(4) 掌握使用 Exception 类捕获异常的方法。

2. 任务描述

编写一个除法运算器，在程序中要处理除数为 0 的异常情况。

3. 实施步骤

1) 创建包

在项目 Lab10 中创建包 task2。

2) 创建文件并进行编辑

在包 task2 中创建 Java 类文件 DivByZero.java，修改文件的内容如下：

```java
package task2;
import java.util.InputMismatchException;
import java.util.Scanner;
/**
 * 处理除数为0的异常
 * @author sf
 */
public class DivByZero {
    public static void main(String[] args) {
        Scanner input=new Scanner(System.in);              //输入对象
        int x,y;
        System.out.print("请输入除数x: ");
        try{
            x=input.nextInt();
            y=10/x;
            System.out.println("10/x 的结果为: "+y);
        }catch(InputMismatchException ex){                 //异常捕获①
            System.out.println("输入的除数数据必须是整数类型");
        }catch(ArithmeticException ex){                    //异常捕获②
            System.out.println("除数不能为0。");
        } catch(Exception ex){                             //其他类型的异常处理
            System.out.println("其他类型的错误。");
        }
        input.close();                                     //关闭输入对象
    }
}
```

3) 运行程序

程序的 3 个运行实例效果如图 10-2 所示。

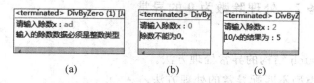

图 10-2　除数为 0 时异常处理的 3 个运行实例效果

10.3.3　任务 3　处理数组的下标越界异常

1. 任务目的

(1) 掌握数组下标越界异常 ArrayIndexOutOfBoundsException 的使用方法。

(2) 掌握通用异常处理类 Exception 的使用方法。

2. 任务描述

给定一个字符串数据，将这个数组中的内容输出，处理数组下标越界异常。

3. 实施步骤

1) 创建包

在项目 Lab10 中创建包 task3。

2) 创建文件并进行编辑

在包 task3 中创建 Java 类文件 ArrayIndexOutOfBound.java，修改文件的内容如下：

```java
package task3;
/**
 * 数组下标越界的异常处理
 * @author sf
 */
public class ArrayIndexOutOfBound {
    public static void main(String[] args) {
        String[] nameArr={"张三","李四","王五","刘六"};
        try{
            for(int i=0;i<=4;i++){
                System.out.println(nameArr[i]);
            }
        }catch(ArrayIndexOutOfBoundsException ex){
            System.out.println("数组的下标越界。");
        } catch(Exception ex){
            System.out.println("其他的异常。");
        }
    }
}
```

3) 运行程序

运行该程序，会看到如图 10-3 所示的运行结果。

图 10-3 数组下标越界异常运行情况

10.3.4 任务 4 特殊字符检查器

1. 任务目的

(1) 掌握 throw 抛出异常的方法。

(2) 掌握 throws 抛出异常的方法。

(3) 掌握自定义异常的定义与使用方法。

2. 任务描述

编写一个程序，对给定的字符串内容进行检查，如果这个字符串的内容全是数字或英文字母，则显示这个字符串，否则抛出异常，提示有非法字符。检查字符是否满足题目的要求时，要按 ASCII 码表中的字符的 ASCII 码值去进行比较检查，数字 0~9 的 ASCII 码值为 48~57，大写字母 A~Z 的 ASCII 码值为 65~90，小写字母的 ASCII 码值为 97~122。

3. 实施步骤

1) 创建包

在项目 Lab10 中创建包 task4。

2）创建自定义异常类并进行编辑

在包 task4 中创建 Java 类异常文件 MyStrChkException.java，修改文件的内容如下：

```java
package task4;
/**
 * 自定义异常类
 * @author sf
 */
public class MyStrChkException extends Exception {        //继承 Exception 类
    private static final long serialVersionUID=1L;        //类的序列化号
    private String content;
    public MyStrChkException(String content) {            //构造方法
        this.content=content;
    }
    public String getContent() {                          //获取描述方法
        return content;
    }
}
```

3）创建异常测试类并进行编辑

在包 task4 中创建 Java 测试类文件 MyStrChkTest.java，修改文件的内容如下：

```java
package task4;
/**
 * 自定义异常测试类
 * @author sf
 */
public class MyStrChkTest {
    /**检查字符串中是否含有非法字符的方法
     * @param str 要检查的字符串
     * @throws MyStrChkException 抛出自定义的异常
     */
    public static void chkStr(String str) throws MyStrChkException{
        char[] array=str.toCharArray();
        int len=array.length;
        for(int i=0;i<len-1;i++){
            //数字 0~9 的 ASCII 码值为 48~57,大写字母 A~Z 的 ASCII 码值为 65~90
            //小写字母的 ASCII 码值为 97~122
            if(!((array[i]>=48 && array[i]<=57)||(array[i]>=65 &&
                array[i]<=90)||(array[i]>=97 && array[i]<=122))){
                throw new MyStrChkException("字符串："+str+"中含有非法字符。");
            }
        }
    }

    public static void main(String[] args) {
```

```
        String str1="abczA09Z";
        String str2="ab!334@";
        try{
            chkStr(str1);
            System.out.println("字符串 1 为："+str1);
            chkStr(str2);
            System.out.println("字符串 2 为："+str2);
        }catch(MyStrChkException ex){
            System.out.println("触发自定义的异常,异常内容如下：");
            System.out.println(ex.getContent());
        }
    }
}
```

4）运行程序

运行测试类 MyStrChkTest，将会看到如图 10-4 所示的运行结果。

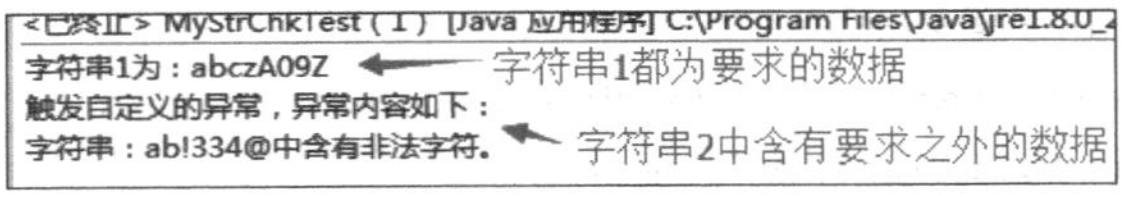

图 10-4　自定义异常进行特殊字符检查器的运行结果

4. 任务扩展

(1) 在此基础上，完成对敏感词汇的过滤。

(2) 将本程序与文件或数据库结构结合，完成自定义敏感词汇信息的过滤。

10.3.5　任务 5　使用 try-with-resource 进行读取文件处理

1. 任务目的

(1) 掌握 throws 抛出异常的方法。

(2) 掌握 try-with-resource 自动释放资源的方法。

(3) 掌握 I/O 异常的处理方法。

2. 任务描述

使用 try-with-resource 读取 C:/mytest.txt 文件的内容，并把内容显示出来，在读取文件过程中，注意关闭读取流对象，并处理 I/O 异常。

3. 实施步骤

1）创建文本文件

在 C 盘根目录下创建文本文件 mytest.txt，在文件中随意输入一些内容即可。

2）创建包

在项目 Lab10 中创建包 task5。

3）创建读取文本文件的程序

在包 task5 中创建读取文本文件的 Java 程序 MyTryWithResource.java，修改该文件的内容如下：

```java
package task5;

import java.io.BufferedReader;
import java.io.FileReader;
import java.io.IOException;

/**
 * 自动释放资源的 try 块
 * @author sf
 */
public class MyTryWithResource {
    public static void main(String[] args) throws IOException {    //抛出 I/O 异常
        //使用 try-with-resource 处理文件读取流
        try (BufferedReader reader=new BufferedReader(new FileReader
            ("c:/mytest.txt"))) {
            StringBuilder builder=new StringBuilder();
            String line=null;
            while ((line=reader.readLine())!=null) {
                builder.append(line);
                builder.append(String.format("%n"));
            }
            System.out.println(builder.toString());              //将内容在控制中显示
        }
    }
}
```

4) 运行程序

运行 MyTryWithResource 类,将会看到如图 10-5 所示的运行结果。

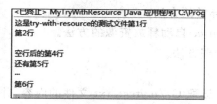

图 10-5 使用 try-with-resource 读取文件程序的运行结果

第 11 章　图形用户界面设计

11.1　实验目的

(1) 掌握图形用户界面编程的步骤。
(2) 掌握常用布局管理类的使用。
(3) 理解容器的嵌套。
(4) 掌握常用组件的使用。
(5) 掌握事件处理的要素。
(6) 掌握 Java 中事件处理的方式。

11.2　实验任务

(1) 任务 1：公司员工信息录入程序。
(2) 任务 2：小学生习题训练程序。
(3) 任务 3："我所喜爱的主食和副食"问卷调查。
(4) 任务 4：员工信息处理菜单。
(5) 任务 5：商场收银软件。

11.3　实验内容

11.3.1　任务 1　公司员工信息录入程序

1. 任务目的

(1) 掌握图形用户界面编程的步骤。
(2) 掌握常用布局管理类的使用。
(3) 基本 ActionEvent 事件的基本概念、事件的产生和事件处理的过程。

2. 任务描述

定义公司的职员信息类，成员变量包括 ID(身份证)、name(姓名)、sex(性别)、birthday(生日)、home(籍贯)、address(居住地)和 number(职员号)，设计一个录入或显示职工信息的程序界面，要求以 FlowLayout 布局安排组件，如图 11-1 所示。当单击"操作"按钮时，输入下一个职工的信息；当单击"退出"按钮时，结束信息的输入，退出程序。

3. 实施步骤

1) UI 布局设计

容器：JFrame。

组件：JLabel 标签、JTextField 文本框、JButton 按钮。

图 11-1 职工信息录入显示程序

布局：FlowLayout。

UI 设计部分参考代码如下：

```java
package task1.ui;
import java.awt.FlowLayout;
import javax.swing.JButton;
import javax.swing.JFrame;
import javax.swing.JLabel;
import javax.swing.JTextField;
public class EmployeeInfoFrame extends JFrame {
    JTextField ID=new JTextField(18);
    JTextField name=new JTextField(10);
    JTextField birthday=new JTextField(10);
    JTextField sex=new JTextField(2);
    JTextField home=new JTextField(18);
    JTextField address=new JTextField(18);
    JTextField brithday=new JTextField(10);
    JTextField number=new JTextField(5);
    JButton operate=new JButton("操作");
    JButton exit=new JButton("退出");
    public EmployeeInfoFrame() {
        super("公司职员信息");
        this.setLayout(new FlowLayout());
        this.add(new JLabel("身份证号码"));
        this.add(ID);
        this.add(new JLabel("姓名"));
        this.add(name);
        this.add(new JLabel("性别"));
        this.add(sex);
        this.add(new JLabel("出生日期"));
        this.add(birthday);
        this.add(new JLabel("籍贯"));
        this.add(home);
        this.add(new JLabel("居住地"));
        this.add(address);
        this.add(new JLabel("职工号"));
        this.add(number);
        this.add(operate);
```

```
        this.add(exit);
        this.setSize(530,300);
        this.setVisible(true);
    }
    public static void main(String[] args) {
        new EmployeeInfoFrame();
    }
}
```

2)事件处理分析

事件源:按钮组件 operate 和 exit。

触发的事件类型:ActionEvent。

事件处理的主线为 ActionEvent→ActionListener→addActionListener。

3)operate 按钮事件处理

编写事件处理类:

```
//内部类
class OperateHandler implements ActionListener{
    @Override
    public void actionPerformed(ActionEvent event) {
        /*保存职工信息处理代码
         * 保存完成
         */
        //清空文本框的录入
        ID.setText("");
        name.setText("");
        ⋮
        operate.setText("下一个");
    }
}
//注册监听
operate.addActionListener(new OperateHandler());
```

4)exit 按钮事件处理

```
//采用匿名内部类的方式
exit.addActionListener(new ActionListener() {

    @Override
    public void actionPerformed(ActionEvent event) {
        System.exit(0);
    }
});
```

5)运行程序

观察结果,如图 11-1 所示。

4. 任务拓展

（1）能不能通过一个事件处理类来处理两个按钮的单击事件呢？

提示：ActionEvent 类封装了两个方法，即 getActionCommand() 和 getSource()。

（2）使用 GridLayout 布局修改以上界面，完成后如图 11-2 所示。

11.3.2 任务 2 小学生习题训练程序

1. 任务目的

（1）掌握容器的嵌套使用。

（2）掌握 KeyEvent 事件的基本概念、事件的产生和事件处理的过程。

2. 任务描述

编程实现一个如图 11-3 所示的小学生习题训练用户界面。并当输入当前题的应答后，按 Enter 键或单击"下一题"按钮进入下一题的应答。要求运算数是 100 以内的数，可视运算数的大小确定某种运算（＋、－、＊、/）。在完成测试之后显示测试结果即测试题目数、答对的题目数和使用时间数（以分计）。

图 11-2　使用 GridLayout 布局

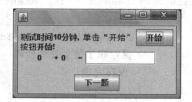
图 11-3　小学生习题训练用户界面

3. 实施步骤

1）UI 布局设计

根据要求该界面有 3 行组成：第一行由一个标签（JLabel）和一个按钮（JButton）组件构成；第二行由 4 个标签（分别表示两个运算数、运算符号和等号）和一个文本框（JTextField）组件构成；第 3 行由一个按钮构成。

每一行的组件添加到 FlowLayout 流布局的 JPanel 容器中，外层的 JPanel 采用 GridLayout 布局按 3 行 1 列的形式将所有的子容器添加到上面。

然后再将封装好的 JPanel 容器添加到 JFrame 上。

UI 设计部分参考代码如下：

```
/*小学生训练题初始界面
 *StudentScreen.java
 */
class StudentScreen extends JPanel {
    JLabel num1=new JLabel("0    ");          //显示第 1 个运算数
    JLabel num2=new JLabel("0    ");          //显示第 2 个运算数
    JLabel op=new JLabel("+");                //显示运算符号
    JLabel info=new JLabel("测试时间 10 分钟,单击"开始"按钮开始!");
    JTextField answer=new JTextField(10);     //输入运算结果
```

```java
        JButton next=new JButton("下一题");         //操作按钮
        JButton start=new JButton("开始");          //操作按钮
        JPanel pan1=new JPanel();
        JPanel pan2=new JPanel();
        JPanel pan3=new JPanel();
        public StudentScreen() {
            setLayout(new GridLayout(3,1));
            //将第1行的组件放在pan1容器中
            pan1.add(info);
            pan1.add(start);
            //将第2行的组件放在pan2容器中
            pan2.add(num1);
            pan2.add(op);
            pan2.add(num2);
            pan2.add(new JLabel("="));
            pan2.add(answer);
            pan3.add(next);
            add(pan1);                              //将pan1(第1行的组件)摆放在界面上
            add(pan2);                              //将pan2(第2行的组件)摆放在界面上
            add(pan3);                              ////pan3(第3行的组件)摆放在界面上
        }
    }
    /**
     * 最外层容器
     */
    public class StudentTrainFrame extends JFrame {
        public StudentTrainFrame() {
            add(new StudentScreen());
            setVisible(true);
            pack();                                 //以最紧凑的方式排列组件
            //设置关闭时退出程序
            setDefaultCloseOperation(JFrame.EXIT_ON_CLOSE);
        }
        public static void main(String[] args) {
            new StudentTrainFrame();
        }
    }
```

2) 事件处理分析

要实现 StudentScreen 界面上的按钮功能,由任务 1 可知,只要实现 ActionListener 接口即可。要处理按键功能,由于其触发的是 KeyEvent 事件,则需要实现 KeyListener 接口,在接口方法中完成按键功能的实现。

根据题意,测试题有两个运算数,运算数可以取 100 以内的随机整数,运算操作包括+、一、*、/四种运算,可以根据两个操作数的大小来确定运算符。因此可以加入一个成员方法 setOperator()来确定运算符。

在答题后按 Enter 键和单击"下一题"按钮执行的是同一个操作,就是完成当前题目答案的核对处理,产生下一题的运算数和运算符,进入下一题的应答。所以加入一个成员方法 compute()来完成该操作,可以在按键方法 keyPressed()、动作按钮方法 actionPerformed()中调用 compute()方法以避免代码重复。

由于在事件处理过程中,会涉及初始界面的变化,保持初始界面不变,在本程序实现时单独定义了一个类 Exercises 去继承 StudentScreen 类,该类负责完成界面的更新和事件处理。

参考代码如下:

```java
/*
 * 小学生训练题 Exercises.java
 */
class Exercises extends StudentScreen implements ActionListener,KeyListener
{
    int count=0;                              //用于记录已答题的数量
    int n1=0,n2=0;                            //两个运算数
    int total=0;                              //记录题目的总数量
    int right=0;                              //记录答对题目的数量
    long timenum=0;                           //记录答题的开始时间
    Random rand=new Random();                 //用于产生随机数
    public Exercises(int total) {
        this.total=total;
        answer.setEnabled(false);             //在没开始答题之前,不得答题
        start.addActionListener(this);
        next.addActionListener(this);
        answer.addKeyListener(this);
    }
    /**
     * ActionListener 接口方法的实现
     */
    public void actionPerformed(ActionEvent e) {
        if (e.getSource()==start) {
            info.setText("共"+total+"题!");
            start.setEnabled(false);          //使开始按钮失效
            answer.setEnabled(true);          //开始答题
            answer.requestFocus();            //获取焦点
            count=right=0;                    //重新置 0
            n1=rand.nextInt(100);
            n2=rand.nextInt(100);
            num1.setText(""+n1);
            num2.setText(""+n2);
            timenum=System.currentTimeMillis(); //获取开始答题的时间
        } else if (e.getSource()==next) {
            compute();                        //调用 compute()成员方法
        }
```

```java
}
/**
 * KeyListener 接口方法的实现
 */
public void keyPressed(KeyEvent e)              //当按一个键时调用它
{
    if (e.getKeyCode()!=e.VK_ENTER)
        return;                                 //如果不是按了 Enter 键,则不处理
    compute();                                  //调用 compute()成员方法
}
public void keyReleased(KeyEvent ke)            //当一个键被释放时调用它
{
    /* 当需要时,输入相关处理代码 */
}
public void keyTyped(KeyEvent ke)               //当输入一个字符键时调用它
{
    /* 当需要时,输入相关处理代码 */
}
/********** 成员方法设置运算符 ***********/
public void setOperator()                       //设置运算符方法
{
    if (n1>50 && n2>50)
        if (n1<n2)
            op.setText("+");                    //n1>50,n2>50,并且 n1<n2 进行加法运算
        else
            op.setText("-");                    //n1>n2 进行减法运算
    else if (n1>50)
        if (n2>10)
            op.setText("-");                    //n1>50,n2>10 进行减法运算
        else
            op.setText("/");                    //n1>50,n2<10 进行除法运算
    else if (n2>50)
        if (n1>10)
            op.setText("+");                    //n2>50,n1>10 进行加法运算
        else
            op.setText(" * ");                  //n2>50,n1≤10 进行乘法运算
    else if (n1>n2 && n2<10)
        op.setText("/");                        //n1>n2,n2<10 进行除法运算
    else if (n1<10 || n2<10)
        op.setText(" * ");                      //n1≤10 或 n2≤10 进行乘法运算
    else
        op.setText("+");                        //其他进行加法运算
}
/********** 成员方法:运算及答案处理 ***********/
public void compute() {
```

```java
        float x=0;                              //定义变量
        if (op.getText().equals("+"))
            x=n1+n2;
        else if (op.getText().equals("-"))
            x=n1-n2;
        else if (op.getText().equals("*"))
            x=n1 * n2;
        else if (op.getText().equals("/"))
            x=n1/n2;
        if (x==Float.parseFloat(answer.getText()))
            right++;
        count++;
        if (count==total)                       //测试结束
        {
            JOptionPane.showMessageDialog(null,"总题数"+total+";答对"+right+
                "道,费时"+(System.currentTimeMillis()-timenum)/60000+"分!");
            System.exit(0);                     //退出程序
        }
        n1=rand.nextInt(100)+1;                 //产生下一题
        n2=rand.nextInt(100)+1;
        num1.setText(""+n1);
        num2.setText(""+n2);
        setOperator();                          //设置运算符号
        answer.setText("");
        answer.requestFocus();
    }
}
```

参考代码说明：

setOperator()用于确定本题的运算符号,其规则如下。

(1) 当两个运算数 n1、n2 均大于 50 时,若 n1>n2 进行加法运算,否则进行减法运算。

(2) 当 n1>50,n2≤50 时,若 n2>10 进行减法运算,否则(n2≤10)进行除法运算。

(3) 当 n2>50,n1≤50 时,若 n1>10 进行加法运算,否则(n1≤10)进行乘法运算。

(4) 当 n1≤50,n2≤50 时,若 n1>n2 与 n2<10 进行除法运算,否则若 n1<10 或 n2<10 时,进行乘法运算,其他进行加法运算。

compute()主要完成以下几个功能。

(1) 根据运算符对两个运算数进行运算获得正确答案。

(2) 将测试者给出的答案和正确答案相比较,若一致 right 计数加 1。

(3) 答题计数 count 加 1。

(4) 将题目总数 total 和答题计数比较,若相等答题结束,显示结果,程序结束。

(5) 产生下一题。

在 KeyListener 接口中有 3 个方法,实际只用了 keyPressed()方法,在方法中只需要对 Enter 键进行处理,其他键不需要考虑。另外两个方法没有用到,所以没有功能实现的程序代码,是两个空方法。

3) 程序运行

单击"开始"按钮后,由初始界面变成如图 11-4 所示的界面。

单击"下一题"按钮或者按 Enter 键后进入如图 11-5 所示的界面。

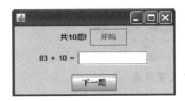
图 11-4　单击"开始"按钮后进入答题界面

图 11-5　答题过程界面

答题完成后,通过弹出框提示答题结果,如图 11-6 所示。

说明:由于答题速度较快,不到 1 分钟答完了 10 道题,所以以上界面显示用时 0 分钟。读者可以修改程序,如果不到一分钟显示以秒计。

4. 任务拓展

(1) 键盘按键事件处理采用匿名内部类或内部类的方式实现。

(2) 用户答题后,如果按 Enter 键或者单击"下一题"按钮,如果答错了,弹出提示框,询问用户是否再想想,如果用户选择是重新做该题。

图 11-6　答题结果界面

11.3.3　任务 3　"我所喜爱的主食和副食"问卷调查

1. 任务目的

(1) 进一步掌握容器的嵌套使用。

(2) 掌握 Java 中鼠标事件(MouseEvent)及其他事件的基本概念,事件的产生和事件处理的基本过程。

2. 任务描述

职工食堂为了方便职工就餐,进行了一次"我所喜爱的主食和副食"调查,调查表中列出了一些食堂经营的主食和副食类,经营者将调查表发给各位职工,让他们推荐自己喜欢的主食和副食。经营者将根据调查结果确定其经营方向。试创建一个调查表式的选项界面,如图 11-7 所示,当鼠标落在选项上时,改变该选项标识的颜色;当鼠标离开选项时恢复其原来的颜色。

3. 实施步骤

1) UI 布局设计

参考本章任务 2,仍然采用布局嵌套的方式来布局界面。

UI 设计部分参考代码如下:

```
import javax.swing.JFrame;
import java.awt.*;
```

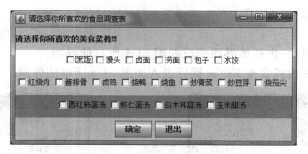

图 11-7　调查表式的选项界面

```java
import javax.swing.*;
public class MealPoll extends JFrame {
    Checkbox[] sele1=new Checkbox[6];
    Checkbox[] sele2=new Checkbox[8];
    Checkbox[] sele3=new Checkbox[4];
    JPanel panel1=new JPanel();
    JPanel panel2=new JPanel();
    JPanel panel3=new JPanel();
    JPanel panel4=new JPanel();
    JButton ok=new JButton("确定");
    JButton exit=new JButton("退出");
    JLabel prompt=new JLabel("请选择你所喜欢的美食菜肴!!!");

    MealPoll() {
        super("请选择你所喜欢的食品调查表");          //调用父类构造器,设置标题栏
        sele1[0]=new Checkbox("米饭");              //加入主食选择菜单项
        sele1[1]=new Checkbox("馒头");
        sele1[2]=new Checkbox("卤面");
        sele1[3]=new Checkbox("涝面");
        sele1[4]=new Checkbox("包子");
        sele1[5]=new Checkbox("水饺");
        sele2[0]=new Checkbox("红烧肉");            //加入副食菜类选择菜单项
        sele2[1]=new Checkbox("酱排骨");
        sele2[2]=new Checkbox("卤鸡");
        sele2[3]=new Checkbox("烧鸭");
        sele2[4]=new Checkbox("烧鱼");
        sele2[5]=new Checkbox("炒青菜");
        sele2[6]=new Checkbox("炒豆芽");
        sele2[7]=new Checkbox("烧茄尖");
        sele3[0]=new Checkbox("西红柿蛋汤");        //加入副食汤类选择菜单项
        sele3[1]=new Checkbox("虾仁蛋汤");
        sele3[2]=new Checkbox("白木耳甜汤");
        sele3[3]=new Checkbox("玉米甜汤");
        panel1.setBackground(Color.white);          //设置容器panel1背景
```

```
            for (int i=0; i<sele1.length; i++)
                panel1.add(sele1[i]);              //将主食选项加入到panel1容器中
            panel3.setBackground(Color.pink);      //设置容器panel2背景
            for (int i=0; i<sele2.length; i++)
                panel2.add(sele2[i]);              //将菜类选项加入到panel2容器中
            for (int i=0; i<sele3.length; i++)
                panel3.add(sele3[i]);              //将汤类选项加入到panel3容器中
            panel4.add(ok);                        //将按钮项加入到panel4容器中
            panel4.add(exit);
            Container pane=this.getContentPane();  //获得界面容器
            pane.setLayout(new GridLayout(5,1));   //界面容器布局5行1列
            pane.add(prompt);                      //将标签项放入界面容器
            pane.add(panel1);                      //将panel1放入界面容器
            pane.add(panel2);                      //将panel2放入界面容器
            pane.add(panel3);                      //将panel3放入界面容器
            pane.add(panel4);                      //将panel4放入界面容器
            this.pack();
            this.setVisible(true);                 //窗口上的内容是可见的
            this.setDefaultCloseOperation(EXIT_ON_CLOSE);
        }
        public static void main(String[] args) {
            new MealPoll();
        }
    }
```

2)事件处理分析

根据题意,要实现监视鼠标位置,用到鼠标操作所产生的 MouseEvent 事件,由于在 MouseListener 接口中提供了 5 个方法,所以实现 MouseListener 接口需要编写这 5 个方法,但事实上,本示例只需要实现两个方法 MouseEntered(鼠标进入)和 MouseExited(鼠标离开),因此可以创建一个 MouseAdapter 适配器类派生的内部类 MouseHandler,重写 MouseEntered()和 MouseExited()方法,进行鼠标进入选项改变其标识颜色,离开时恢复其颜色的处理。

鼠标适配器类的实现参考代码如下:

```
//鼠标事件适配器类的实现
class MouseHandler extends MouseAdapter {
    public void mouseEntered(MouseEvent event)           //鼠标进入事件
    {
        Checkbox checkBox= (Checkbox) event.getSource(); //获得事件源对象
        oldColor=checkBox.getForeground();               //记录对象的原前景色
        checkBox.setForeground(Color.BLUE);              //设置对象的前景色为蓝色
    }
    public void mouseExited(MouseEvent event)            //鼠标离开事件
    {
```

```
        Checkbox checkBox= (Checkbox) event.getSource();    //获得事件源对象
        checkBox.setForeground(oldColor);                    //恢复对象的前景色
    }
}
```

注意注册事件监听,主要代码如下:

```
for (int i=0; i<sele1.length; i++){
    panel1.add(sele1[i]);                                   //将主食选项加入到panel1容器中
    sele1[i].addMouseListener(new MouseHandler());
}
```

自己补充完整副食类、汤类的事件监听。

3) 程序运行

鼠标移入时,前景色变为蓝色,如图 11-8 所示,移出时又恢复原来的前景色。

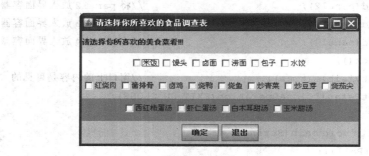

图 11-8　鼠标移入时

4. 任务拓展

给"确定"按钮和"退出"按钮增加事件处理,当单击"确定"按钮时,能够获取用户的选项并通过弹出框提示,当单击"退出"按钮时退出应用程序。

11.3.4　任务 4　员工信息处理菜单

1. 任务目的

(1) 了解菜单栏(JmenuBar)、菜单(Jmenu)和菜单项(JmenuItem)的功能和作用。

(2) 了解如何使用 JmenuBar、Jmenu 和 JMenuItem 类创建菜单。

2. 任务描述

创建一个公司员工基本信息管理的菜单用户界面,如图 11-9 所示。

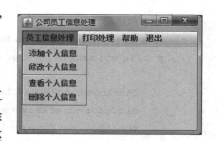

3. 实施步骤

1) 简要分析

一般来说,员工的基本信息管理可以包括登记员工信息、修改员工信息、查看员工信息、在员工调离后删除其信息等,此外还需要打印员工信息表以备案,有时还需要打印某个或某些员工的信息,对有些项目的说明提供帮助功能。

图 11-9　员工信息处理菜单用户界面

使用javax.swing类包中提供的JMenuBar、JMenu、JMenuItem功能将其划分为4个菜单：员工信息处理（添加个人信息、修改个人信息、查看个人信息、删除个人信息）、打印处理（打印所有信息、打印指定信息、打印指定员工的信息）、帮助和退出。

2）参考代码

```java
import javax.swing.JFrame;
import javax.swing.JMenu;
import javax.swing.JMenuBar;
import javax.swing.JMenuItem;

/*这是一个公司职员信息管理的菜单界面
 *程序的名字是 EmployeeMenu.prg
 */
public class EmployeeMenu extends JFrame {
    JMenuBar empBar=new JMenuBar();                         //定义菜单栏对象
    JMenu mess=new JMenu("员工信息处理");                     //定义员工信息处理菜单对象
    JMenuItem addMess=new JMenuItem("添加个人信息");         //定义菜单项对象
    JMenuItem editMess=new JMenuItem("修改个人信息");
    JMenuItem checkMess=new JMenuItem("查看个人信息");
    JMenuItem delMess=new JMenuItem("删除个人信息");
    JMenu prtMess=new JMenu("打印处理");                     //定义打印处理菜单对象
    JMenuItem prt_all=new JMenuItem("打印所有信息");         //定义菜单项对象
    JMenuItem prt_part=new JMenuItem("打印指定信息");
    JMenuItem prt_one=new JMenuItem("打印指定员工的信息");
    JMenu help=new JMenu("帮助");                            //定义帮助菜单对象
    JMenuItem info=new JMenuItem("关于帮助");                //定义菜单项对象
    JMenuItem subject=new JMenuItem("帮助主题");
    JMenu exit=new JMenu("退出");                            //定义结束菜单对象

    public EmployeeMenu()                                   //构造方法
    {
        this.setTitle("公司员工信息处理");                    //设置框架窗体标题
        /*以下把文件菜单项加入到Mess菜单中*/
        mess.add(addMess);
        mess.add(editMess);
        mess.addSeparator();                                //添加分割条
        mess.add(checkMess);
        mess.add(delMess);
        /*以下把编辑菜单项加入到prtMess菜单中*/
        prtMess.add(prt_all);
        prtMess.addSeparator();                             //添加分割条
        prtMess.add(prt_part);
        prtMess.addSeparator();                             //添加分割条
        prtMess.add(prt_one);
        /*以下把帮助菜单项加入到Help菜单中*/
```

```
            help.add(info);
            help.add(subject);
            /*以下把所有菜单加入到菜单栏中*/
            empBar.add(mess);
            empBar.add(prtMess);
            empBar.add(help);
            empBar.add(exit);
            this.setJMenuBar(empBar);        //将菜单栏加入框架窗口
            this.setSize(300,200);
            this.setVisible(true);
            this.setDefaultCloseOperation(3);
        }
        public static void main(String[] args) {
            new EmployeeMenu();
        }
    }
```

3）程序运行

观察结果,如图 11-9 所示。

4. 任务拓展

通过查阅 Java 的 API 文档,试着给菜单加上事件处理功能。

11.3.5 任务5 商场收银软件

1. 任务目的

（1）进一步掌握容器的嵌套使用。

（2）掌握视图部分和业务逻辑部分的分离。

2. 任务描述

实现如图 11-10 所示的商场收银软件。

3. 实施步骤

1）UI 布局设计

考虑到中间区域最大及整体的结构,最外层采用边界布局 BorderLayout,中间区域和南部区域都只放置一个组件,只有北部区域放置了很多组件,所以必然得容器嵌套。

本例将使用盒式容器 BoxLayout,最外层是创建一个水平方向的 BoxLayout,里面 3 个垂直方向上的 BoxLayout。

确定了布局后,我们再来分析一下涉及的组件有哪些。

标签：JLabel。

按钮：JButton。

文本框：JTextField。

图 11-10 商场收银软件

文本域：JTextArea。

下拉列表：JComboBox。

小技巧：如何查找到自己所需要的组件呢？在 Java API 文档中，找到 JComponent，看看它的子类，由于组件命名都非常规范，相信你能很轻松地找到自己所需要的组件。

这里面，就是 JComboBox 组件稍微复杂些，我们一起来看看如何使用。

先看看如何构建它。在 Java API 中找到如下的构造方法：

JComboBox(Object[] items)：创建包含指定数组中的元素的 JComboBox。

相信你很快就可以写出如下代码：

```java
String[] items={ "正常收费","打8折","打7折","打5折","满300返100"};
JComboBox method=new JComboBox(items);
```

UI 设计部分参考代码如下：

```java
package task;
import java.awt.Dimension;
import java.awt.Font;
import java.awt.event.ActionEvent;
import java.awt.event.ActionListener;
import javax.swing.Box;
import javax.swing.BoxLayout;
import javax.swing.JButton;
import javax.swing.JComboBox;
import javax.swing.JFrame;
import javax.swing.JLabel;
import javax.swing.JPanel;
import javax.swing.JScrollPane;
import javax.swing.JTextArea;
import javax.swing.JTextField;
public class CashSystem extends JFrame {
    public static final int WIDTH=400;
    public static final int HEIGHT=400;
    private JTextField price;
    private JTextField number;
    private JButton confirm;
    private JButton cancel;
    private JTextArea showArea;
    private JScrollPane jShowArea;
    private JLabel showResult;
    private double total=0.0;
    private JComboBox method;
    public void startFrame() {
        this.setTitle("商场收银系统");
        //屏幕居中
        Dimension d=this.getToolkit().getScreenSize();
```

```java
this.setLocation((d.width-WIDTH)/2,(d.height-HEIGHT)/2);
this.setSize(WIDTH,HEIGHT);
//关闭时退出当前程序
this.setDefaultCloseOperation(JFrame.EXIT_ON_CLOSE);
//设置大小不可调整
this.setResizable(false);
//北部第1个垂直方向的容器
JPanel v1=new JPanel();
BoxLayout b1=new BoxLayout(v1,BoxLayout.Y_AXIS);
v1.setLayout(b1);
v1.add(new JLabel("单价:"));
v1.add(new JLabel("数量:"));
v1.add(new JLabel("计算方式:"));
//北部第2个垂直方向的容器
JPanel v2=new JPanel();
BoxLayout b2=new BoxLayout(v2,BoxLayout.Y_AXIS);    //产生一个容器
v2.setLayout(b2);
price=new JTextField(15);
number=new JTextField(15);
String[] items={ "正常收费","打8折","打7折","打5折","满300返100" };
method=new JComboBox(items);
v2.add(price);
v2.add(number);
v2.add(method);
//北部第3个垂直方向的容器
JPanel v3=new JPanel();
BoxLayout b3=new BoxLayout(v3,BoxLayout.Y_AXIS);    //产生一个容器
v3.setLayout(b3);
confirm=new JButton("确定");
cancel=new JButton("重置");
v3.add(confirm);
v3.add(cancel);
//北部外层水平方向的容器
JPanel h=new JPanel();
BoxLayout b4=new BoxLayout(h,BoxLayout.X_AXIS);    //产生一个容器
h.setLayout(b4);
h.add(v1);
h.add(v2);
h.add(v3);
//中部区域的文本域组件
showArea=new JTextArea(6,15);
//给文本域加上滚动条,内容显示不开时自动出现滚动条
jShowArea=new JScrollPane(showArea);
//南部区域放置的组件
showResult=new JLabel("显示总价");
```

```java
        //确定按钮事件处理
        confirm.addActionListener(new ConfirmHandler());
        this.add(h,"North");
        this.add(jShowArea,"Center");
        this.add(showResult,"South");
        this.setVisible(true);
    }
    public static void main(String[] args) {
        CashSystem cs=new CashSystem();
        cs.startFrame();
    }
}
```

2) 编写业务逻辑类

为了实现代码复用,将业务逻辑部分单独封装成一个类 Cash。
Cash 类的实现参考代码如下:

```java
package task.update.model;
public class Cash {
    private double price;
    private int num;
    public Cash(double price,int num) {
        super();
        this.price=price;
        this.num=num;
    }
    //打 8 折、7 折、5 折的计算方法
    public double getCash(double rate) {
        return price * num * rate;
    }
    //满 300 返 100 的计算方法
    public double getCash(int moneyCondition,int moneyReturn) {
        double money;
        money=price * num;
        if (money>moneyCondition) {
            return money- (int) money/moneyCondition * moneyReturn;
        } else
            return money;
    }
}
```

3)"重置"按钮事件处理

采用匿名内部类的方式,参考代码如下:

```java
cancel.addActionListener(new ActionListener() {
    public void actionPerformed(ActionEvent e) {
        price.setText(null);
```

```
            number.setText(null);
            showArea.setText(null);
            showResult.setText(null);
        }
    });
```

4)"确定"按钮事件处理

采用内部类的方式,参考代码如下:

```
class ConfirmHandler implements ActionListener {
    double totalPrice=0;
    public void actionPerformed(ActionEvent e) {
        Cash cash=new Cash(Double.parseDouble(price.getText()),
                Integer.parseInt(number.getText()));
        String condition="正常收费";
        //判断选择的计算方式,下标从 0 开始
        switch(method.getSelectedIndex()) {
        case 0:
            totalPrice=cash.getCash(1);
            condition="正常收费";
            break;
        case 1:
            totalPrice=cash.getCash(0.8);
            condition="打 8 折";
            break;
        case 2:
            totalPrice=cash.getCash(0.7);
            condition="打 7 折";
            break;
        case 3:
            totalPrice=cash.getCash(0.5);
            condition="打 5 折";
            break;
        case 4:
            totalPrice=cash.getCash(300,100);
            condition="满 300 返 100";
            break;
        }
        total=total+totalPrice;
        showArea.append(condition+"--单价:"+price.getText()+"数量:"
                +number.getText()+"合计:"+totalPrice+"\n");
        //设置标签显示的字体
        showResult.setFont(new Font("楷体",Font.ITALIC,20));
        showResult.setText("总计:"+total+"元");
    }
}
```

5）程序运行

观察结果,如图 11-10 所示。

4. 任务拓展

以上代码实现了业务层和视图层的分离,可是每当商场有新的活动时,还是需要修改 Cash 类的实现,根据前面所学习的面向对象的相关知识,能不能进一步地改进程序呢?

11.4 课后巩固练习

（1）设计一个如图 11-11(a)所示的系统注册界面。用户输入姓名、性别、生日、爱好、电邮、学历信息,然后单击"注册"按钮,则会弹出如图 11-11(b)所示的系统注册成功界面,并显示该用户的注册信息。

(a)　　　　　　　　　　(b)

图 11-11　注册界面

（2）设计一个如图 11-12 所示的计算器界面,并实现连续加、减、乘、除的功能。

图 11-12　计算器界面

第 12 章 输入输出流

12.1 实验目的

（1）掌握 File 类的使用。
（2）掌握字符流 Reader 和 Writer 的使用。
（3）掌握字节流 InputStream 和 OutputStream 的使用。
（4）理解字节流和字符流的使用场合。
（5）掌握内存操作流。
（6）掌握打印流。
（7）理解缓冲流。
（8）了解 Scanner 类实现文件操作。
（9）理解对象序列化。

12.2 实验任务

（1）任务 1：FileWriter 和 BufferedWriter 比较。
（2）任务 2：给源程序加入行号。
（3）任务 3：统计英语短文字母 A 出现的次数。
（4）任务 4：简易 Java 考试系统。

12.3 实验内容

12.3.1 任务 1 FileWriter 和 BufferedWriter 比较

1. 任务目的

（1）掌握字符流的使用。
（2）理解缓冲流的作用。

2. 任务描述

分别使用 FileWriter 和 BufferedWriter 往文件中写入 10 万个随机数，比较用时的多少。

3. 实施步骤

（1）直接使用 FileWriter 往文件中写入 10 万个随机数。
提示：计算用时使用 System.currentTimeMills 求时间差。
参考代码如下：

```
import java.io.FileWriter;
```

```
import java.io.IOException;
public class Task1_1 {
    public static void main(String[] args) throws IOException {
        long start=System.currentTimeMillis();          //获取开始时间
        //创建字符文件输出流
        FileWriter fw=new FileWriter("D:\\task1.txt");
        for (int i=1; i<=100000; i++) {
            fw.write((int) (Math.random() * 10000)+" \t");
        }
        fw.close();                                     //别忘了关闭流
        long t=System.currentTimeMillis()-start;        //计算时间差
        System.out.println("Time elapsed: "+t+"ms");
    }
}
```

(2) 运行程序，输出结果：Time elapsed：93ms。D 盘 task1. txt 文件中写入的内容如图 12-1 所示。

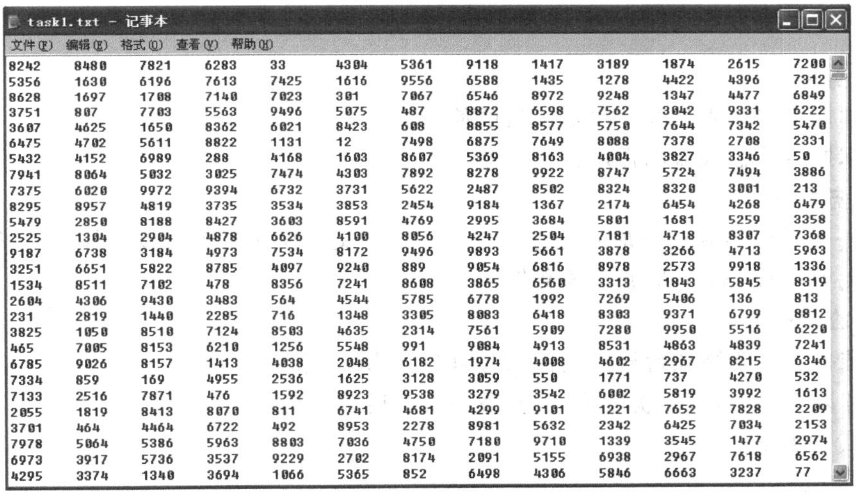

8242	8480	7821	6283	33	4304	5361	9118	1417	3189	1874	2615	7200
5356	1630	6196	7613	7425	1616	9556	6588	1435	1278	4422	4396	7312
8628	1697	1708	7140	7023	301	7067	6546	8972	9248	1347	4477	6849
3751	807	7703	5563	9496	5075	487	8872	6598	7562	3042	9331	6222
3607	4625	1650	8362	6021	8423	608	8855	8577	5750	7644	7342	5470
6475	4702	5611	8822	1131	12	7498	6875	7649	8088	7378	2708	2331
5432	4152	6989	288	4168	1603	8607	5369	8163	4004	3827	3346	50
7941	8064	5032	3025	7474	4303	7892	8278	9922	8747	5724	7494	3886
7375	6020	9972	9394	6732	3731	5622	2487	8502	8324	8320	3001	213
8295	8957	4819	3735	3534	3853	2454	9184	1367	2174	6454	4268	6479
5479	2850	8188	8427	3603	8591	4769	2995	3684	5801	1681	5259	3358
2525	1304	2904	4878	6626	4100	8056	4247	2504	7181	4718	8307	7368
9187	6738	3184	4973	7534	8172	9496	9893	5661	3878	3266	4713	5963
3251	6651	5822	8785	4097	9240	889	9054	6816	8978	2573	9918	1336
1534	8511	7102	478	8356	7241	8608	3865	6560	3313	1843	5845	8319
2604	4306	9430	3483	564	4544	5785	6778	1992	7269	5406	136	813
231	2819	1440	2285	716	1348	3305	8083	6418	8303	9371	6799	8812
3825	1050	8510	7124	8503	4635	2314	7561	5909	7280	9950	5516	6220
465	7005	8153	6210	1256	5548	991	9084	4913	8531	4863	4839	7241
6785	9026	8157	1413	4038	2048	6182	1974	4008	4602	2967	8215	6346
7334	859	169	4955	2536	1625	3128	3059	550	1771	737	4270	532
7133	2516	7871	476	1592	8923	9538	3279	3542	6002	5819	3992	1613
2055	1819	8413	8070	811	6741	4681	4299	9101	1221	7652	7828	2209
3701	464	4464	6722	492	8953	2278	8981	5632	2342	6425	7034	2153
7978	5064	5386	5963	8803	7036	4750	7180	9710	1339	3545	1477	2974
6973	3917	5736	3537	9229	2702	8174	2091	5155	6938	2967	7618	6562
4295	3374	1340	3694	1066	5365	852	6498	4306	5846	6663	3237	77

图 12-1 D 盘 task1.txt 文件中写入的内容

(3) 使用 BufferedWriter 往文件中写入 10 万个随机数。

参考代码如下：

```
import java.io.BufferedWriter;
import java.io.FileWriter;
import java.io.IOException;
public class Task1_2 {
    public static void main(String[] args) throws IOException {
        long start=System.currentTimeMillis();
        //使用 BufferedWriter 装饰 FileWriter 类,使其具有缓冲的功能
        BufferedWriter fw=new BufferedWriter(new FileWriter("D:\\task1.txt"));
        for (int i=1; i<=100000; i++) {
```

```
            fw.write((int) (Math.random() * 10000)+" \t");
        }
        fw.close();
        long t=System.currentTimeMillis()-start;
        System.out.println("Time elapsed: "+t+"ms");
    }
}
```

(4) 运行程序。

运行结果如下：

Time elapsed: 78ms

4. 任务拓展
(1) 直接使用 FileReader 读取 task1.txt 文件的内容。
(2) 使用 BufferedReader 读取 task1.txt 文件的内容。
(3) 比较两种读文件方法的用法区别。

12.3.2 任务2 给源程序加入行号

1. 任务目的
(1) 掌握字符输入流和字符输出流的使用。
(2) 掌握缓冲输入流的常用方法。

2. 任务描述
编写一个可以给源程序加入行号的程序。利用文件输入流读入该文件，加入行号后，将这个文件另存为以 txt 为扩展名的文件中。

3. 实施步骤
(1) 准备好需要添加行号的文件，如读取 Task1_1.java 文件。
(2) 读取文件内容，每次读取一行，添加行号后，输出到另外一个文件中。

参考代码如下：

```
import java.io.BufferedReader;
import java.io.BufferedWriter;
import java.io.File;
import java.io.FileReader;
import java.io.FileWriter;
import java.io.IOException;
public class Task2 {
    public static void main(String[] args) throws IOException {
        File inFile=new File("D:","task1_1.java");         //添加当前文件路径
        File outFile=new File("D:\\temp.txt");             //目标文件
        //创建指向 inFile 的输入流
        FileReader fileReader=new FileReader(inFile);
        BufferedReader bufferedReader=new BufferedReader(fileReader);
        //创建指向文件 tempFile 的输出流
```

```
        FileWriter tofile=new FileWriter(outFile);
        BufferedWriter out=new BufferedWriter(tofile);
        int i=0;                                        //记录行号
        String s=bufferedReader.readLine();             //从源文件读取一行
        while (s!=null) {
            i++;
            out.write(i+" "+s);
            out.newLine();                              //换行
            s=bufferedReader.readLine();                //从源文件中读取下一行
        }
        fileReader.close();
        bufferedReader.close();
        out.flush();
        out.close();
        tofile.close();
    }
}
```

(3) 运行程序。

观察结果,在 D 盘生成了一 temp.txt 文件,文件内容如图 12-2 所示。

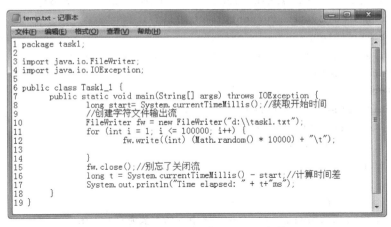

图 12-2　加入行号的 temp.txt 文件

4. 任务拓展

利用 JFileChooser 类提示用户选择一个源文件。

13.3.3　任务 3　统计英语短文字母 A 出现的次数

1. 任务目的

(1) 掌握字节输入流的使用。

(2) 掌握字节输出流的使用。

2. 任务描述

在文本文件 poem.txt 中包含有很长篇幅的英语短文,编写程序要求统计文件的所有短文中包含英文字母 A(不区分大小写)的个数,并显示统计的时间。

3. 实施步骤

1) 准备好 poem.txt 文件

poem.txt 文件中的内容如图 12-3 所示。

```
If you think you are beaten, you are;
If you think you dare not, you don't;
If you want to win but think you can't;
It's almost a cinch you won't.
If you think you'll lose, you're lost;
For out of the world we find Success begins with a fellow's will;
It's all in a state of mind.
Life's battles don't always go To the stronger and faster man,
But sooner or later the man who wins Is the man who thinks he can.
```

图 12-3 poem.txt 文件中的内容

2) 使用 FileInputStream 读取文件中的内容并统计字母 A 出现的次数

参考代码如下:

```java
import java.io.FileInputStream;
import java.io.IOException;
public class Task3 {
    public static void main(String[] args) throws IOException {
        long time=System.currentTimeMillis();
        String filename="poem.txt";
        FileInputStream f=new FileInputStream(filename);
        int count=0;
        int c;
        while ((c=f.read())!=-1) {      //读不到字符时返回-1
            if (c=='A'||c=='a') {
                count++;
            }
        }
        f.close();
        System.out.println("poem.txt 中 A 的个数为:"+count);
        time=System.currentTimeMillis()-time;
        System.out.println("统计 A 的时间为:"+time);
    }
}
```

3) 程序运行

运行结果如下:

poem.txt 文件中 A 的个数为:22
统计 A 的时间为:0

4. 任务拓展

使用缓冲流 BufferedInputStream 来读取文件中的内容,并比较所用时间。

13.3.4 任务 4　简易 Java 考试系统

1. 任务目的

（1）掌握如何进行分层设计程序。
（2）掌握文件输入流和输出流的使用。
（3）掌握对象输入流和输出流的使用。
（4）能够灵活地运用输入输出流解决实际的应用。
（5）进一步掌握图形用户界面的设计及其事件处理。

2. 任务描述

该系统可以生成 10 道选择题,每次生成的顺序都不同,单击题号按钮,显示相应的题目进行作答,都答完了,单击"关闭"按钮会有一个提示框,单击"是"按钮,在弹出的对话框中输入自己学号后两位,保存本次答题结果,如图 12-4～图 12-7 所示。

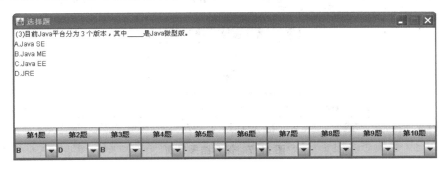

图 12-4　答题过程界面

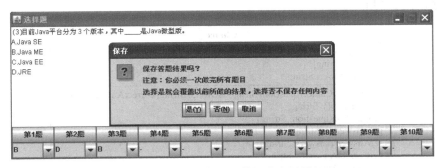

图 12-5　单击"关闭"按钮弹出对话框

3. 实施步骤

1）思路分析

这里我们不过多地分析界面设计部分,主要给大家分析涉及 I/O 操作的部分。

问题 1：题目从哪里得到？

我们借助于文件保存 10 道题目,包括问题、答案和正确答案,为了便于拆分字符串,以 ♯ 号作为分隔符,文件中的题目按照如下格式进行组织：

编译 Java 程序的命令是_____。♯java♯javadoc♯javac♯cd♯javac

问题 2：如何获取文件中的所有题目？

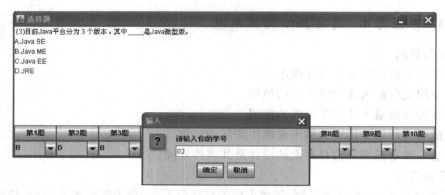

图 12-6 单击"是"按钮弹出输入学号对话框

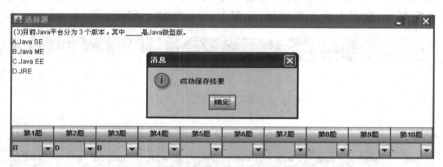

图 12-7 保存结果界面

这里采用面向对象的编程思想,首先将题目单独地抽出一个类来描述,单选题 Choice 类的设计如图 12-8 所示。

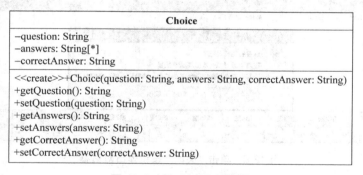

图 12-8 Choice UML 类图

属性 question 存放问题,answers 数组存放 4 个可选答案,correctAnswer 存放正确答案。

接下来就可以构造 ChoiceDao 类,里面封装方法 getAllChoices(),核心代码如下:

```
public class ChoiceDao {
    List<Choice>choices=new ArrayList<Choice>();    //存放所有单选题
        public List<Choice>getAllChoices() {
        //读取文件中的题目
```

```
        return choices;
    }
}
```

如何读取文件中的题目呢?

本例中要一次读取一行,然后对其进行拆分,所以我们借助于缓冲流 BufferedReader 来实现,核心代码如下:

```
//创建文件字符输入流
Reader reader=new FileReader(new File("choicetest.txt"));
//使用 BufferedReader 装饰 FileReader,使之带有缓冲功能
BufferedReader bufferedReader=new BufferedReader(reader);
String choiceString=null;
String[] choice=new String[6];                          //存放拆分后的题目
//使用 while 循环每次读取文件的一行
while ((choiceString=bufferedReader.readLine())!=null) {
    //调用 String 类的 split()方法拆分字符串,将拆分后的结果保存在 choice 数组中
    choice=choiceString.split("#");
    //打乱可选答案的顺序
    //添加单选题到 choices 集合
}
```

如何打乱可选答案的顺序呢?

拆分后,choice[0]存放的问题,choice[5]存放的是正确答案,我们要打乱的是 choice[1] 至 choice[4],借助于 Random 类,随机生成 1~4 的下标,核心代码如下:

```
int[] randomNumber=new int[4];                          //存放随机生成的下标 1、2、3、4
Random random=new Random();
for (int i=0; i<randomNumber.length; i++) {
    //随机生成 1~4 之间的一个随机数赋给当前值
    randomNumber[i]=random.nextInt(4)+1;
    //如果随机生成的数和之前生成的数有相同的,则重新生成 i--一次,然后再 i++,还是当前的 i
    for (int j=0; j<=i-1; j++) {
        if (randomNumber[i]==randomNumber[j]) {
            i--;
            break;
        }
    }
}
```

问题 3:学生答完题后如何保留学生的学号和答题成绩呢?

采用面向对象的编程思想,将抽出学生实体类,其 UML 类图如图 12-9 所示。

注意:该实体类要实现 Serializable 接口,实现学生对象的持久化保存,借助于对象输出流来实现,在本例中抽出了一个通用的对象保存于读取的类 ObjectDao,里面封装了两个方法 saveObject()和 readObject(),StudentDao 继承 ObjectDao,就可以直接使用其定义的方法。

Student
−number: String −score: int
<<create>>+Student(number: String, score: int) +getNumber(): String +setNumber(number: String) +getScore(): int +setScore(score: int) +toString(): String

图 12-9　Student 的 UML 类图

2）根据上面的分析，编写实体类

单项选择题实体类参考代码如下：

```java
/**
 * 单项选择题实体类
 */
public class Choice {
    private String question;                              //问题
    private String[] answers;                             //4个可选答案
    private String correctAnswer;                         //正确答案
    public Choice(String question,String[] answers,String correctAnswer) {
        super();
        this.question=question;
        this.answers=answers;
        this.correctAnswer=correctAnswer;
    }
    public String getQuestion() {
        return question;
    }
    public void setQuestion(String question) {
        this.question=question;
    }
    public String[] getAnswers() {
        return answers;
    }
    public void setAnswers(String[] answers) {
        this.answers=answers;
    }
    public String getCorrectAnswer() {
        return correctAnswer;
    }
    public void setCorrectAnswer(String correctAnswer) {
        this.correctAnswer=correctAnswer;
    }
}
```

学生实体类参考代码如下：

```java
import java.io.Serializable;
/**
 * 学生实体类
 */
public class Student implements Serializable{
    private String number;                              //学号
    private int score;                                  //得分
    public Student(String number,int score) {
        super();
        this.number=number;
        this.score=score;
    }
    public String getNumber() {
        return number;
    }
    public void setNumber(String number) {
        this.number=number;
    }
    public int getScore() {
        return score;
    }
    public void setScore(int score) {
        this.score=score;
    }
    public String toString() {
        return "Student [number="+number+",score="+score+"]";
    }
}
```

3）根据上面的分析，编写业务逻辑类

ChoiceDao 类主要是完成对单选题的业务操作，实现代码参考如下：

```java
/**
 * 单选题操作类
 */
public class ChoiceDao {
    List<Choice>choices=new ArrayList<Choice>();
    int[] randomNumber=new int[4];                      //存放随机生成的下标 1、2、3、4
    public List<Choice>getAllChoices() {
        //存放所有单选题
        try {
            readFromFile();
        } catch (FileNotFoundException e) {
            e.printStackTrace();
```

```java
        } catch (IOException e) {
            e.printStackTrace();
        }
        return choices;
    }
    private void readFromFile() throws FileNotFoundException,IOException{
        String[] choice=new String[6];                          //存放拆分后的题目
        Reader reader=new FileReader(new File("choicetest.txt"));
        BufferedReader bufferedReader=new BufferedReader(reader);
        String choiceString=null;
        while ((choiceString=bufferedReader.readLine())!=null) {
            //拆分字符串
            choice=choiceString.split("#");
            //随机生成下标1、2、3、4,打乱顺序
            shuffle();
            //添加单选题
            choices.add(new Choice(choice[0],new String[] {
                choice[randomNumber[0]],choice[randomNumber[1]],
                choice[randomNumber[2]],choice[randomNumber[3]]},
                choice[choice.length-1]));
        }
    }
    //打乱题目顺序
    private void shuffle() {
        int[] randomNumber=new int[4];                  //存放随机生成的下标1、2、3、4
        Random random=new Random();
        for (int i=0; i<randomNumber.length; i++) {
            //随机生成1~4之间的一个随机数赋给当前值
            randomNumber[i]=random.nextInt(4)+1;
            //如果随机生成的数和之前生成的数有相同的,则重新生成i--一次,然后再i++,还
            //是当前的i
            for (int j=0; j<=i-1; j++) {
                if (randomNumber[i]==randomNumber[j]) {
                    i--;
                    break;
                }
            }
        }
    }
}
```

StudentDao类主要是完成判断学生答题是否正确判分的,参考代码如下:

```
/**
 * 学生操作类
 */
```

```java
public class StudentDao extends ObjectDao {
    public static void saveStudent(Student student,String fileName) {
        saveObject(student,fileName);
    }
    public static Student readStudent(String fileName) {
        return (Student) readObject(fileName);
    }
    //判断用户回答是否正确
    public static int getScore(Choice choice,String answer) {
        int score=0;
        String correctAnswer=choice.getCorrectAnswer();
        if ("A".equals(answer)) {
            score=judge(choice,0,correctAnswer);
        } else if ("B".equals(answer)) {
            score=judge(choice,1,correctAnswer);
        } else if ("C".equals(answer)) {
            score=judge(choice,2,correctAnswer);
        } else if ("D".equals(answer)) {
            score=judge(choice,3,correctAnswer);
        }
        return score;
    }
    private static int judge(Choice choice,int i,String correctAnswer) {
        String userAnswer=choice.getAnswers()[i];
        if (correctAnswer.equals(userAnswer)) {
            return 2;
        }
        return 0;
    }
}
```

ObjectDao 类主要是使用对象输入输出流完成对象的保存和读取操作,参考代码如下:

```java
/**对象保存和读取通用类**/
public class ObjectDao {
    //保存对象
    public static void saveObject(Object object,String fileName) {
        ObjectOutputStream oos=null;
        try {
            FileOutputStream fos=new FileOutputStream(new File(fileName+".dat"));
            oos=new ObjectOutputStream(fos);
            oos.writeObject(object);
        } catch (FileNotFoundException e) {
            e.printStackTrace();
        } catch (IOException e) {
            e.printStackTrace();
```

```java
        } finally {
            try {
                oos.close();
            } catch (IOException e) {
                e.printStackTrace();
            }
        }
    }
    //读取对象
    public static Object readObject(String fileName) {
        ObjectInputStream ois=null;
        Object object=null;
        try {
            FileInputStream fis=new FileInputStream(new File(fileName));
            ois=new ObjectInputStream(fis);
            object=ois.readObject();
        } catch (FileNotFoundException e) {
            e.printStackTrace();
        } catch (IOException e) {
            e.printStackTrace();
        } catch (ClassNotFoundException e) {
            e.printStackTrace();
        } finally {
            try {
                ois.close();
            } catch (IOException e) {
                e.printStackTrace();
            }
        }
        return object;
    }
}
```

4）UI 设计及其事件处理

参考代码如下：

```java
public class ChoiceWindow extends JFrame {
    private List<Choice>choicesList=new ArrayList<Choice>();
    private JButton[] tihao=new JButton[10];
    private JComboBox[] choices=new JComboBox[10];
    private ChoiceDao choiceDao=new ChoiceDao();
    private Choice choice=null;;
    int j=0;
    private int score;
    public static void main(String[] args) {
        new ChoiceWindow("选择题");
```

```java
    }
    public ChoiceWindow(String title) {
        super(title);
        //设置窗口居中显示
        Toolkit tk=Toolkit.getDefaultToolkit();
        Dimension d=tk.getScreenSize();
        setSize(d.width/2,d.height/2);
        setLocation(d.width/4,d.height/4);
        //设置窗口大小不可调整
        setResizable(false);
        //创建10行15列的文本域组件
        final JTextArea choiceArea=new JTextArea(10,15);
        choiceArea.setText("请选择题号,然后在相应题号下面作答。");
        //设置不可编辑
        choiceArea.setEditable(false);
        //添加到JFrame容器的中央区域
        this.add(choiceArea);
        //创建2行1列的中间容器
        JPanel timu=new JPanel();
        timu.setLayout(new GridLayout(2,1));
        //创建1行10列的子容器存放题号按钮
        JPanel tihaoPanel=new JPanel();
        tihaoPanel.setLayout(new GridLayout(1,10));
        //创建1行10列的子容器存放选项下拉列表
        JPanel choicePanel=new JPanel();
        choicePanel.setLayout(new GridLayout(1,10));
        //获取所有的单选题
        choicesList=choiceDao.getAllChoices();
        for (int i=0; i<tihao.length; i++) {
            //创建10个按钮组件并添加到tihaoPanel
            tihao[i]=new JButton("第"+(i+1)+"题");
            tihaoPanel.add(tihao[i]);
            //注册监听
            tihao[i].addActionListener(new ActionListener() {
                @Override
                public void actionPerformed(ActionEvent e) {
                    choiceArea.setText(" ");
                    //取得所单击按钮的题号,第10题特殊处理
                    if (e.getActionCommand().equals("第10题")) {
                        j=9;
                        choice=choicesList.get(9);
                    } else {
                        //比如第2题,e.getActionCommand().charAt(1)取出的是字符'2',
                        //49是'1','2'-'1'得出应该取得下标是1
                        j=e.getActionCommand().charAt(1)-49;
```

```java
                    choice=choicesList.get(e.getActionCommand().charAt(1)-49);
                }
                //在 choiceArea 区域追加问题
                choiceArea.append(choice.getQuestion()+"\n");
                //在 choiceArea 区域追加 4 个选项
            choiceArea.append("A."+choice.getAnswers()[0]+"\n");
            choiceArea.append("B."+choice.getAnswers()[1]+"\n");
            choiceArea.append("C."+choice.getAnswers()[2]+"\n");
            choiceArea.append("D."+choice.getAnswers()[3]+"\n");
            }
        });
        //创建下拉列表选项并添加到 choicePanel
        choices[i]=new JComboBox(new String[] { "-","A","B","C","D" });
        choicePanel.add(choices[i]);
    }
    timu.add(tihaoPanel);
    timu.add(choicePanel);
    //添加 timu 面板至南部区域
    this.add(timu,"South");
    //紧凑显示
    this.pack();
    //设置单击"关闭"按钮时什么都不做
    this.setDefaultCloseOperation(JFrame.DO_NOTHING_ON_CLOSE);
    setVisible(true);
}
//处理窗口事件
@Override
protected void processWindowEvent(WindowEvent e) {
    super.processWindowEvent(e);
    //如果窗口正在关闭
    if (e.getID()==WindowEvent.WINDOW_CLOSING) {
        //弹出确认对话框
        int result=JOptionPane.showConfirmDialog(this,
                "保存答题结果吗?\n 注意:你必须一次做完所有题目\n 选择是就会覆盖以
                    前所做的结果,选择否不保存任何内容",
                "保存",JOptionPane.YES_NO_CANCEL_OPTION);
        if (result==JOptionPane.YES_OPTION) {
            String number=JOptionPane.showInputDialog("请输入你的学号");
            //计算成绩
            computeScore();
            Student student=new Student(number,score);
            //保存结果
            StudentDao.saveStudent(student,number);
        JOptionPane.showMessageDialog(this,"成功保存结果","消息",
                JOptionPane.INFORMATION_MESSAGE);
```

```java
                System.exit(1);
            } else if (result==JOptionPane.NO_OPTION) {
                JOptionPane.showMessageDialog(this,"结果不保存","消息",
                        JOptionPane.INFORMATION_MESSAGE);
                System.exit(1);
            } else if (result==JOptionPane.CANCEL_OPTION) {
            }
        } else {
            //忽略其他事件,交给 JFrame 处理
            super.processWindowEvent(e);
        }
    }
    //计算得分
    public void computeScore() {
        for (int i=0; i<choices.length; i++) {
            Choice choice=choicesList.get(i);
            //获得下拉列表用户选择的答案
            String answer= (String) choices[i].getSelectedItem();
            score+=StudentDao.getScore(choice,answer);
        }
    }
}
```

4. 任务拓展

本程序实现了单项选择题的考试判分,大家能不能给这个系统加入填空题呢?

第 13 章　Java 集合框架

13.1　实验目的

（1）掌握 List 接口的特点及其使用场合。
（2）掌握 Set 接口的特点及其使用场合。
（3）理解引用相等性和对象相等性。
（4）掌握 Map 接口的特点和使用场合。
（5）理解泛型。

13.2　实验任务

（1）任务 1：使用 List 模拟图书系统实现歌曲的增、删、改、查。
（2）任务 2：使用 Map 模拟电话号码管理程序。

13.3　实验内容

13.3.1　任务 1　使用 List 模拟图书系统实现歌曲的增、删、改、查

1. 任务目的

（1）掌握 List 接口的特点及其使用场合。
（2）掌握 Set 接口的特点及其使用场合，并与 Set 接口进行比较。
（3）理解引用相等性和对象相等性。
（4）理解泛型。

2. 任务描述

图书馆里有各种各样的图书，儿童书和计算机图书琳琅满目，要求设计这样一个系统能够实现添加图书、删除图书、修改图书和查询图书的功能。

3. 实施步骤

（1）面向对象分析和设计。
先思考如下两个问题。
① 根据问题描述，你能抽象出哪些对象呢？
② 对象映射成的类应该包含哪些属性和方法呢？
（2）根据分析定义类 Book，该类为抽象类，封装图书所具有的基本属性。
参考代码如下：

```
//抽象的父类 Book
abstract class Book {
```

```java
//书的 ISBN 号、名称、价格、信息
private String ISBN;
private String name;
private float price;
private String info;
public Book(String iSBN,String name,float price,String info) {
    super();
    ISBN=iSBN;
    this.name=name;
    this.price=price;
    this.info=info;
}
public String getISBN() {
    return ISBN;
}
public String getName() {
    return name;
}
public float getPrice() {
    return price;
}

public String getInfo() {
    return info;
}
}
```

(3) 定义 Book 类的子类 ChildBook 和 ComputerBook。

参考代码如下：

```java
//儿童书
class ChildBook extends Book{
    public ChildBook(String iSBN,String name,float price,String info) {
        super(iSBN,name,price,info);
    }
    public String toString(){
        return "儿童书 书名："+this.getName()+"价格"+this.getPrice()+"介绍"+
            this.getInfo();
    }
}
//计算机书
class ComputerBook extends Book{
    public ComputerBook(String iSBN,String name,float price,String info) {
        super(iSBN,name,price,info);
    }
    public String toString(){
```

```
        return "计算机书 书名："+this.getName()+"价格"+this.getPrice()+"介绍"+
            this.getInfo();
    }

}
```

(4) 根据分析定义类 BookDao，即图书操作类，负责对书籍的增、删、改、查操作。
思考如下问题：
① 如何使用 List 保存所有的书籍？
② 如何添加一本图书？
③ 如何删除一本图书？
④ 如何修改一本图书？
⑤ 如何查询一本图书？
⑥ 如何输出所有书籍？
参考代码如下：

```
public class BookDao {
    private List<Book>allBooks;                    //存放所有的图书
    public BookDao() {
        super();
        allBooks=new ArrayList<Book>();
    }
    //查询所有的书籍
    public List<Book>getAllBooks() {
        return allBooks;
    }
    //增加书籍
    public void addBook(Book book){
        this.allBooks.add(book);
    }
    //删除书籍
    public void removeBook(Book book){
        this.allBooks.remove(book);
    }
    //查询书籍,根据书名查询某本书
    public Book getBookByName(String name){
        Book book=null;
        for(Book tempbook:this.allBooks){
            if(tempbook.getName().equals(name)){
                book=tempbook;
                break;
            }
        }
        return book;
    }
```

```java
        //模糊查询
        public List<Book> index(String keyword){
            List<Book> list=new ArrayList<Book>();
            for(Book tempbook:this.allBooks){
                if(tempbook.getName().indexOf(keyword)!=-1){
                    list.add(tempbook);
                }
            }
            return list;
        }
        //列出所有的书籍
        public void showAllBooks() {
            getAllBooks();
            Iterator<Book> iter=allBooks.iterator();
            while (iter.hasNext()) {
                Book b=iter.next();
                System.out.println(b);
            }
        }
}
```

如何修改图书,实现将图书按照书名的方式进行排序的方法。

提示:排序相关的接口 Comparable 或 Comparator。

(5) 定义测试驱动类,模拟添加几本图书,删除图书,修改图书,查询图书的过程。

参考代码如下:

```java
public class Task1 {
    public static void main(String[] args) {
        Book b1=new ChildBook("200401","一千零一夜",10.0f,"一些传说故事");
        Book b2=new ChildBook("200402","小鸡吃大灰狼",20.0f,"一件奇怪的事");
        Book b3=new ChildBook("200403","HALIBOTE",25.0f,"魔幻故事");
        Book b4=new ComputerBook("200404","Java",65.0f,"Java语言");
        Book b5=new ComputerBook("200405","C++",50.0f,"C++语言");
        Book b6=new ComputerBook("200406","Linux",50.0f,"服务器搭建");
        BookDao bookDao=new BookDao();
        bookDao.addBook(b1);
        bookDao.addBook(b2);
        bookDao.addBook(b3);
        bookDao.addBook(b4);
        bookDao.addBook(b5);
        bookDao.addBook(b6);
        //假设将C++这本书删掉
        Book deletedBook=bookDao.getBookByName("C++");       //先进行查找
        bookDao.removeBook(deletedBook);                     //删除掉查到的这本书
        System.out.println("删除后");
        bookDao.showAllBooks();                              //查询并列出所有的书籍
```

        }
    }

(6) 运行程序,观察结果。

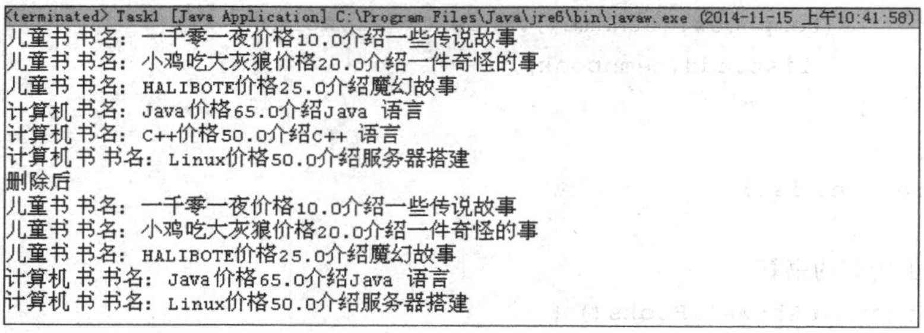

```
<terminated> Task1 [Java Application] C:\Program Files\Java\jre6\bin\javaw.exe (2014-11-15 上午10:41:58)
儿童书 书名:  一千零一夜价格10.0介绍一些传说故事
儿童书 书名:  小鸡吃大灰狼价格20.0介绍一件奇怪的事
儿童书 书名:  HALIBOTE价格25.0介绍魔幻故事
计算机 书名:  Java价格65.0介绍Java 语言
计算机 书名:  C++价格50.0介绍C++ 语言
计算机 书名:  Linux价格50.0介绍服务器搭建
删除后
儿童书 书名:  一千零一夜价格10.0介绍一些传说故事
儿童书 书名:  小鸡吃大灰狼价格20.0介绍一件奇怪的事
儿童书 书名:  HALIBOTE价格25.0介绍魔幻故事
计算机 书名:  Java价格65.0介绍Java 语言
计算机 书名:  Linux价格50.0介绍服务器搭建
```

图 13-1 任务 1 运行结果图

以上程序,若添加同一本书,能否添加成功?如果系统不允许添加同一本书的话,应该怎么修改程序呢?

(7) 使用 Set 接口来重写程序,并比较 List 接口和 Set 接口的特点及使用场合。

4. 任务拓展

(1) 实现按作者名进行排序。

(2) 增加命令行菜单选择。

13.3.2 任务 2 使用 Map 模拟电话号码管理程序

1. 任务目的

(1) 掌握 Map 接口的特点和使用场合。

(2) 进一步理解泛型。

2. 任务描述

开发一款电话号码管理程序,这个程序具有电话号码的添加、删除、修改和查询功能;在添加操作时,如果存在重复名,则显示提示信息,并提示用户输入另外的名称,如果输入的姓名不在记录中,将提示用户是否添加这个电话号码记录。

3. 实施步骤

1) 设计文件来保存电话号码信息。

在项目文件夹下建立 phonebook.txt 文件,内容如下:

```
zhangsan/13475089008
lisi/15254305055
wangwu/13475089006
```

2) 定义 PhoneBook 类封装增、删、改、查等方法

考虑到这是很典型的键值对,即 name→phone,选择使用 Map 接口,又因为我们希望按人名排序,所以最终选择 TreeMap 集合来存放读出来的数据。在 PhoneBook 类中定义 phones 成员来存放读取到的电话信息,此处仍然用到泛型,代码如下:

```
private Map<String,String>phones=new TreeMap<String,String>();
```

定义方法 readPhoneBooks() 读取文件中的所有电话信息,方法如下:

```java
private void readPhoneBooks() throws IOException {
    File file=new File("phonebook.txt");
    BufferedReader reader=new BufferedReader(new FileReader(file));
    String line=null;
    while ((line=reader.readLine())!=null) {
        String[] tokens=line.split("/");     //split()方法会用反斜线来拆分电话信息
          phones.put(tokens[0],tokens[1]);   //调用 put()方法存放名字和电话
    }
}
```

在 PhoneBook 类里面定义方法 displayAllPhones(),用于显示读取到的所有电话信息,方法如下:

```java
public void displayAllphones() {
    //得到键的集合,即名字集合
    Set<String>keySet=phones.keySet();
    //使用 StringBuffer 来拼接字符串
    StringBuffer stringBuffer=new StringBuffer("你的电话本上有如下记录:\n");
    //遍历 set 集合,根据名字取得电话
    for(String name:keySet){
        String phone=phones.get(name);
            stringBuffer.append(name+"-->"+phone+"\n");
    }
    JOptionPane.showMessageDialog(null,stringBuffer.toString());
}
```

在 PhoneBook 类里面定义方法 addNewPhone() 添加一条电话记录,此方法应该能检查是不是已经有同名的用户存在了,如果存在的话应该提示用户重新输入,方法如下:

```java
public void addNewPhone() throws IOException {
    name=JOptionPane.showInputDialog("请输入名字: ");
    phone=JOptionPane.showInputDialog("请输入电话号码: ");
    //调用处理名字重复的方法
    processNameDuplicate();
    phones.put(name,phone);
    //更新文件记录,插入一条新的记录
    updatePhonebook(name,phone);
    String message="添加新的记录成功\n 名字: "+name+"\n 电话号码: "+phone;
    JOptionPane.showMessageDialog(null,message);
}
```

处理名字重复的方法 processNameDuplicate()定义如下:

```java
private void processNameDuplicate() {
    while (phones.containsKey(name)) {     //循环直到不包含重复的名字为止
        String message="在电话本上存在同名的记录\n 请使用另外的名字";
        name=JOptionPane.showInputDialog(message);
```

```
        }
    }
```

更新文件记录的方法如下:

```java
private void updatePhonebook(String name,String phone) throws IOException {
    PrintWriter printWriter=new PrintWriter(new FileWriter(new File(
        "phonebook.txt"),true));            //此处 true 意味着追加到文件末尾
    printWriter.println(name+"/"+phone);
    printWriter.close();
}
```

根据姓名查询电话的 search()方法定义如下:

```java
public void search() throws IOException {
    String choice=null;
    name=JOptionPane.showInputDialog("请输入你要查询的名字");
    if (phones.containsKey(name)) {
        phone=phones.get(name);
        display(name,phone);
    } else {
        String message="这个名字在你的电话本中不存在\n 你想添加这条记录吗?(y/n)";
        choice=JOptionPane.showInputDialog(message);
        if (choice.matches("[y|Y]")) {
            addNewPhone();
        }
    }
}
```

以上的 display()方法定义如下:

```java
private void display(String name,String phone) {
    String message="name:"+name+"\n"+"phone:"+phone;
    JOptionPane.showMessageDialog(null,message);
}
```

电话本程序还应该提供给用户一个选择菜单,让用户选择想要对电话本进行的操作。makeChoice()方法定义如下:

```java
public String makeChoice() {
    String choice=null;
    String message="欢迎使用电话本程序…\n"
        +"1 查看所有电话\n"
        +"2 添加电话\n"
        +"3 查询电话\n"+"4 退出\n";
    boolean done=false;
    while (!done) {
        choice=JOptionPane.showInputDialog(message);
        if (choice.matches("[1|2|3|4]"))
```

```
            done=true;
        else
            JOptionPane.showMessageDialog(null,"输入错误");
    }
    return choice;
}
```

3）程序运行

观察结果，如图 13-2～图 13-13 所示。

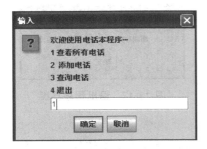

图 13-2 菜单界面

图 13-3 显示所有电话记录

图 13-4 是否继续界面

图 13-5 输入 y 后，再次进入菜单界面

图 13-6 提示输入姓名

图 13-7 提示输入电话号码

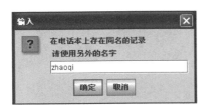

图 13-8 提示输入其他名字

图 13-9 添加新记录成功

图 13-10　查询

图 13-11　如果存在，显示查询结果

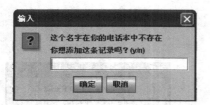
图 13-12　如果不存在，提示是否添加

图 13-13　退出程序界面

4. 任务拓展

改写以上程序，重新进行 GUI 设计，如 4 个菜单选项变成 4 个按钮操作，当单击按钮时进行相应的处理。

第 14 章 Java 网络编程

14.1 实验目的

(1) 掌握 IP 与端口的概念。
(2) 掌握 URL 类及 URLConnection 类、HttpURLConnection 类的用法。
(3) 掌握 InetAddress 类的用法。
(4) 掌握 TCP 与 UDP 的特点和区别。
(5) 掌握 Socket 编程原理及 TCP 通信编程步骤。
(6) 掌握 UDP 编程原理与 UDP 通信编程步骤。

14.2 实验任务

(1) 任务 1：显示 URL 对象的相关属性。
(2) 任务 2：获取本机和远程服务器地址的方法。
(3) 任务 3：检查本机指定范围内的端口是否已经使用。
(4) 任务 4：使用 TCP 通信编写聊天软件。
(5) 任务 5：使用 UDP 通信编写聊天程序。

14.3 实验内容

14.3.1 任务 1 显示 URL 对象的相关属性

1. 任务目的

(1) 掌握 URL 对象的实例化方法。
(2) 掌握 URL 对象的异常处理方法。
(3) 掌握获取 URL 对象各个属性的方法。

2. 任务描述

定义一个 URL 对象，在百度网站的首页中使用查询命令执行"天气预报"的查询。显示查询过程中所用到的 URL 实例对象的相关属性。

3. 实施步骤

1) 创建项目

创建本章的练习项目 Lab14。

2) 创建包

在项目 Lab14 中创建包 task1。

3）创建文件并修改

在包 task1 中创建 Java 类文件 URLBaidu.java，修改该文件的内容如下：

```java
package task1;
import java.io.IOException;
import java.net.MalformedURLException;
import java.net.URL;
public class URLBaidu {
    public static void main(String[] args) {
        try {
            //实例化 URL 对象
            URL url=new URL("http://www.baidu.com/index.htm?sw=天气");
            System.out.println("权限信息："+url.getAuthority());    //权限信息
            try {
                //获取 URL 对象内容
                System.out.println("对象内容："+url.getContent());
            } catch (IOException e) {                              //处理 I/O 异常
                e.printStackTrace();
            }
            //获取默认端口号
            System.out.println("默认端口号："+url.getDefaultPort());
            System.out.println("文件名："+url.getFile());           //获取文件名
            System.out.println("主机名："+url.getHost());           //获取主机名
            System.out.println("URL 路径："+url.getPath());         //获取 URL 路径
            System.out.println("端口号："+url.getPort());           //获取端口号
            System.out.println("协议名："+url.getProtocol());       //获取协议名
            System.out.println("查询信息："+url.getQuery());        //获取查询信息
            System.out.println("URL 锚点："+url.getRef());          //获取 URL 的锚点
            System.out.println("使用者："+url.getUserInfo());       //获取使用者信息
        } catch (MalformedURLException e) {
            e.printStackTrace();
        }
    }
}
```

4）运行程序

程序的运行效果如图 14-1 所示。

4. 任务拓展

（1）使用上面的方法显示查询百度首页时的 URL 对象的属性。

提示：百度首页的地址为 http://www.baidu.com。

（2）使用上面的方法显示查询 CSDN 首页时的 URL 对象的属性。

提示：CSDN 首页的地址为 http://www.csdn.net。

（3）使用上面的方法显示查询任意网站的首页时的 URL 对象的属性。

提示：将上面程序中的 URL 实例化参数替换为其他要查询的网站的 URL 地址即可。

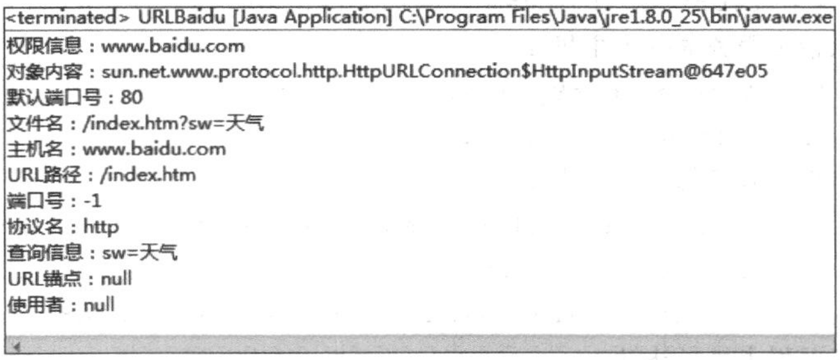

```
<terminated> URLBaidu [Java Application] C:\Program Files\Java\jre1.8.0_25\bin\javaw.exe
权限信息：www.baidu.com
对象内容：sun.net.www.protocol.http.HttpURLConnection$HttpInputStream@647e05
默认端口号：80
文件名：/index.htm?sw=天气
主机名：www.baidu.com
URL路径：/index.htm
端口号：-1
协议名：http
查询信息：sw=天气
URL锚点：null
使用者：null
```

图 14-1　查询天气的 URL 实例对象的属性运行效果

14.3.2　任务 2　获取本机和远程服务器地址的方法

1. 任务目的

(1) 掌握 InetAddress 对象的实例化方法。
(2) 掌握使用 InetAddress 的 getLocalHost()方法获取本机地址的方法。
(3) 掌握使用 InetAddress 的 getByName()方法获取服务器地址的方法。
(4) 掌握使用 InetAddress 的 getAllByName()方法获取所有的服务器地址的方法。

2. 任务描述

使用 InetAddress 类编写一个程序,实现对本机 IP 地址的读取与显示、获取远程主机的一个地址并进行显示、获取远程主机的多个地址并进行显示。

3. 任务分析

按任务描述可知,本任务要解决的问题有 3 个。
(1) 如何获取本机的 IP 地址?
(2) 如何获取远程服务器的地址?
(3) 如何获取远程服务器的多个地址?
要解决上面的 3 个问题,可以使用 InetAddress 类及该类的方法。

4. 实施步骤

1) 创建包

在项目 Lab14 中创建包 task2。

2) 创建文件并进行编辑

在包 task2 中创建 Java 类文件 MyInetAddress.java,修改该类文件的内容如下:

```java
package task2;
import java.io.IOException;
import java.net.InetAddress;
/**
 * 获取本机和远程服务器地址的方法
 * @author sf
 */
public class MyInetAddress {
    public static void main(String[] args) throws IOException {
        InetAddress addr=InetAddress.getLocalHost();        //获取本机 IP
```

```
            System.out.println("local host : "+addr);
            //获取指定服务的一个主机IP
            addr=InetAddress.getByName("baidu.com");
            System.out.println("baidu : "+addr);
            //获取指定服务的所有主机IP
            InetAddress[] addrs=InetAddress.getAllByName("baidu.com");
            for(int i=0 ;i<addrs.length ;i++)
                System.out.println("baidu : "+addrs[i]+" number : "+i);
            //获取远程主机可达性
            System.out.println(InetAddress.getByName("localhost")
                .isReachable(1000));
        }
}
```

3) 运行程序

程序的运行效果如图 14-2 所示。

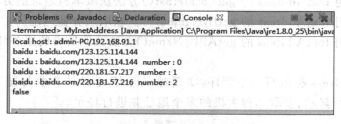

图 14-2 获取本机和远程服务器地址程序的运行效果

5. 任务拓展

(1) 把 baidu.com 换成 www.baidu.com，查看运行效果。

(2) 把程序中的 baidu.com 修改为 google.com，查看运行效果。

(3) 把程序中的 baidu.com 修改为任意的网站的域名或 URL 地址，查看运行效果。

14.3.3 任务3 检查本机指定范围内的端口是否已经使用

1. 任务目的

(1) 掌握 InetAddress 对象的 getLocalHost()方法取得本机地址的方法。

(2) 掌握 Socket 对象的实例化方法。

(3) 掌握 InetAddress 对象的异常处理方法。

(4) 掌握使用 Socket 对象的异常处理方法。

2. 任务描述

编写一个程序，检查本机指定端口范围的端口(如 1～256)是否已经被使用。

3. 实施步骤

1) 创建包

在项目 Lab14 中创建包 task3。

2) 创建文件并进行编辑

在包 task3 中创建 Java 类文件 MyCheckPort.java，修改该类文件的内容如下：

```java
package task3;
import java.io.IOException;
import java.net.InetAddress;
import java.net.Socket;
import java.net.UnknownHostException;
/**
 * 检查某个范围中的端口是否被占用
 * @author sf
 */
public class MyCheckPort {
    public static void main(String[] args) {
        for(int i=1; i<=256; i++){
            try{
                InetAddress local=InetAddress.getLocalHost();
                Socket socket=new Socket(local,i);
                //如果不输出异常,则输出该端口已经被占用
                System.out.println("本机已经使用了端口: "+i);
            }catch(UnknownHostException ex){
                ex.printStackTrace();
            }catch(IOException ex){
                //因为端口被占用时,会大量地抛出这个异常,可以将这个异常注释
                //ex.printStackTrace();
            }
        }
        System.out.println("检查完毕");
    }
}
```

3) 运行程序

运行程序,检查程序的运行效果。

注意:因为要检查的端口范围比较大,创建的对象比较多,因此本程序的运行耗时会较长。

4. 任务拓展

(1) 检查所有的端口(0~65535)是否被占用。

(2) 检查1~20范围中的端口是否被占用,并将异常抛出,查看异常信息。

14.3.4 任务4 使用TCP通信编写聊天软件

1. 任务目的

(1) 理解TCP通信程序的编写及运行过程。

(2) 掌握ServerSocket对象的使用方法。

(3) 掌握Socket对象的使用方法。

(4) 掌握使用输入输出流读写Socket通信信息的方法。

2. 任务描述

使用Socket编程技术,开发一个使用TCP进行网络聊天的软件,该软件包含一个服务

端程序和一个客户端程序,服务端程序启动后,监听客户端的请求,并为客户端提供服务,客户端程序启动,输入要聊的内容,即可以完成与服务端的聊天。

3. 实施步骤

1) 创建包

在项目 Lab14 中创建包 task4。

2) 创建服务端程序文件并进行编辑

在包 task4 中创建服务端 Java 类文件 MyServer.java,修改该类文件的内容如下:

```java
package task4;
import java.io.BufferedReader;
import java.io.InputStreamReader;
import java.io.PrintWriter;
import java.net.ServerSocket;
import java.net.Socket;
import java.util.Scanner;

/**
 * 使用 TCP 与 Socket 编写的聊天服务端
 * @author sf
 */
public class MyServer {
    public static void main(String[] args) {
        try{
            ServerSocket server=null;              //服务端 Socket 对象
            try{
                server=new ServerSocket(8888);     //实例化 Server
                System.out.println("服务端已经准备好,若要退出输入 bye");
            }catch(Exception ex){
                System.out.println("不能创建网络监听器,原因是:"+ex.getMessage());
            }
            Socket socket=null;                    //通信对象
            try{
                socket=server.accept();
            }catch(Exception ex){
                System.out.println("服务出错,原因是:"+ex.getMessage());
            }
            Scanner input=new Scanner(System.in);  //输入实例
            //由 Socket 对象得到输出流,并构造 PrintWriter 对象
            PrintWriter os=new PrintWriter(socket.getOutputStream());
            //取得 Socket 的输入流
            BufferedReader is=new BufferedReader(new InputStreamReader(socket
                        .getInputStream()));
            String words=input.next();    //输入要说的话
            while(!"bye".equals(words)){
                os.println(words);                 //将获取的输入字符串输出到 Server
```

```java
                os.flush();                            //刷新输出流,使服务器端马上接收到该字符串
            System.out.println("服务器: "+words);        //输出客户端的字符串
            System.out.println("客户端: "+is.readLine()); //服务器的字符串
                words=input.next();                    //接收一个新的客户端字符串
            }
            is.close();                                //关闭Socket的输入流
            os.close();                                //关闭Socket的输出流
            input.close();                             //关闭本地的输入
            socket.close();                            //关闭Socket
        }catch(Exception ex){
            ex.printStackTrace();
        }
    }
}
```

3) 创建客户端程序文件并进行编辑

在包 task4 中创建客户端 Java 类文件 MyClient.java,修改该类文件的内容如下:

```java
package task4;
import java.io.BufferedReader;
import java.io.InputStreamReader;
import java.io.PrintWriter;
import java.net.Socket;
import java.util.Scanner;
/**
 * 使用 TCP 与 Socket 编写的聊天客户端
 * @author sf
 */
public class MyClient {
    public static void main(String[] args) {
        try{
            Socket socket=new Socket("127.0.0.1",8888);
            Scanner input=new Scanner(System.in);        //输入实例
            System.out.println("输入你想要说的话,如果要退出的话,输入 bye");
            //由 Socket 对象得到输出流,并构造 PrintWriter 对象
            PrintWriter os=new PrintWriter(socket.getOutputStream());
            //取得 Socket 的输入流
            BufferedReader is=new BufferedReader(new InputStreamReader(socket
                        .getInputStream()));
            String words=input.next();   //输入要说的话
            while(!"bye".equals(words)){
                os.println(words);                //将获取的输入字符串输出到 Server
                os.flush();                       //刷新输出流,使服务器端马上接收到该字符串
            System.out.println("客户端: "+words);   //输出客户端的字符串
            System.out.println("服务器: "+is.readLine()); //服务器的字符串
                words=input.next();                //接收一个新的客户端字符串
```

```
            }
            is.close();              //关闭Socket的输入流
            os.close();              //关闭Socket的输出流
            input.close();           //关闭本地的输入
            socket.close();          //关闭Socket
        }catch(Exception ex){
            ex.printStackTrace();
        }
    }
}
```

4) 运行程序

该聊天软件的运行步骤：启动服务端程序→启动客户端程序→在客户端程序中输入聊天内容后回车→在服务端程序中输入聊天内容后回车，在任何一端不聊天时，输入 bye，则会结束该端的程序，图 14-3 和图 14-4 为该软件的一个运行实例。

图 14-3　聊天软件服务端运行效果　　　　图 14-4　聊天软件客户端运行效果

4．任务拓展

（1）先启动客户端程序，再启动服务端程序，将会出现什么运行结果？

（2）在该程序的基础上，结合 Swing 界面技术，编写简易的桌面 TCP 聊天程序。

14.3.5　任务5　使用 UDP 通信编写聊天程序

1．任务目的

（1）掌握 DatagramSocket 对象的使用方法。

（2）掌握 DatagramPacket 对象的使用方法及数据的发送与接收方法。

（3）掌握 InetAddress 对象的使用方法。

（4）掌握 Socket 的异常处理方法。

（5）掌握 Socket 与 Swing、线程、异常的综合应用方法。

2．任务描述

使用 UDP Socket 技术和 Swing 界面编程技术，编写如图 14-5 所示的 UDP 聊天程序，该界面继承了 JFrame 窗体类，能处理界面文本框中的回车事件和按钮的单击事件，在"发送信息框"中回车或单击了"发送"按钮后，可以将"发送信息框"中的文本信息，发送到左侧文本框中所指定的接收地址中去。

3．任务分析

要实现任务中所指描述的界面功能，就要用到 Swing 编程知识；要使用 UDP 进行通信，就要用到 DatagramSocket 对象的属性和方法；程序中使用 DatagramPacket 来对消息信息进行封装和解析；在解析本机地址和对方的地址时要用到 InetAddress 对象；在编写程序

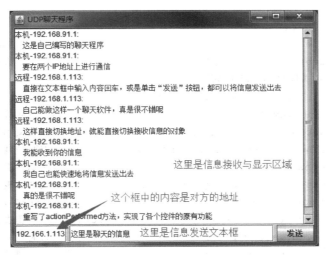

图 14-5 UDP 聊天程序运行界面

时,将对方的地址框属性设置为可以修改,这样就可以在程序运行状态下动态地修改接收地址,从而灵活地改变聊天对象;为了程序使用的方便性,要处理"发送信息"文本框的按键事件、"发送"按钮的单击事件,可以通过重写 actionPerformed() 方法来实现这个功能。

4. 实施步骤

1）创建包

在项目 Lab14 中创建包 task5。

2）创建 UDP 聊天程序文件并进行编辑

在包 task5 中创建 Java 类文件 UDPChat.java,修改该类文件的内容如下：

```
package task5;
import java.awt.BorderLayout;
import java.awt.event.ActionEvent;
import java.awt.event.ActionListener;
import java.io.IOException;
import java.net.*;
import javax.swing.*;

public class UDPMessage extends JFrame implements ActionListener {
    private static final long serialVersionUID=1L;
    private JTextArea text;                    //信息接收文本域
    private JTextField ipText;                 //IP 文本框
    private JTextField sendText;               //信息发送文本框
    private JButton button;                    //发送按钮
    private DatagramSocket socket;             //数据报套接字
    private JScrollBar vsBar;                  //滚动条

    /**
     * 默认的构造方法
```

```java
     */
    public UDPChat() {
        setTitle("UDP聊天程序");                       //设置窗体的标题
        setBounds(100,100,500,400);                     //窗体定位与大小
        //默认的关闭操作为退出程序
        setDefaultCloseOperation(JFrame.EXIT_ON_CLOSE);
        text=new JTextArea();                           //实体信息接收文本域
        text.setEditable(false);                        //信息接收文本域不可以编辑
        JScrollPane textPanel=new JScrollPane(text);    //信息接收文本域添加滚动面板
        vsBar=textPanel.getVerticalScrollBar();         //获取滚动面板的垂直滚动条
        add(textPanel,BorderLayout.CENTER);             //添加滚动面板到窗口居中的位置
        JPanel panel=new JPanel();                      //创建Panel面板
        BorderLayout panelLayout=new BorderLayout();    //创建边界管理器
        panelLayout.setHgap(5);                         //设置布局水平边界
        panel.setLayout(panelLayout);                   //将布局管理器注入到Panel面板
        ipText=new JTextField("192.166.1.100");         //实例化IP文本框
        panel.add(ipText,BorderLayout.WEST);            //添加文本框到Panel面板
        sendText=new JTextField();                      //实例化发送文本框
        sendText.addActionListener(this);               //添加文本框的事件监听器
        panel.add(sendText,BorderLayout.CENTER);        //添加信息文本框到Panel面板
        button=new JButton("发送");                     //实例化发送按钮
        panel.add(button,BorderLayout.EAST);            //添加按钮到Panel面板
        add(panel,BorderLayout.SOUTH);                  //将Panel添加到窗体
        setVisible(true);                               //显示窗体
        server();                                       //调用server()方法
        button.addActionListener(this);                 //添加按钮事件监听器
    }

    /** 服务方法 **/
    private void server(){
        try{
            socket=new DatagramSocket(8888);            //实例化套接字
            byte[] buf=new byte[1024];
            //创建接收数据的数据包
            final DatagramPacket dpl=new DatagramPacket(buf,buf.length);
            Runnable myrun=new Runnable(){              //定义线程
                public void run() {
                    while(true){                        //使用while无限循环体
                        try{
                            Thread.sleep(100);          //线程休眠时间为100ms
                            socket.receive(dpl);        //接收数据包
                            int length=dpl.getLength();
                            //获取数据包的字符串信息
                            String message=new String(dpl.getData(),0,length);
                            //获取IP地址
```

```java
                    String ip=dp1.getAddress().getHostAddress();
                    if(!InetAddress.getLocalHost().getHostAddress().equals(ip)){
                        //显示地址-->换行-->显示信息-->换行
                        text.append("远程-"+ip+":\n "+message+"\n");
                    }
                    vsBar.setValue(vsBar.getMaximum());   //控制信息滚动
                }catch(IOException ex){
                    ex.printStackTrace();
                }catch(InterruptedException ex){
                    ex.printStackTrace();
                }
            }
        };
    };
    new Thread(myrun).start();                      //启动上面所定义的线程
    }catch(SocketException ex){
        ex.printStackTrace();
    }
}

/**
 * 重写界面控件的事件处理,直接接收和处理控件的事件
 */
@Override
public void actionPerformed(ActionEvent arg0) {
    sendMsg();                                      //调用发送信息的方法
}

/**
 * 执行发送信息的方法
 */
private void sendMsg(){
    try{
        String ip=ipText.getText();                 //获取IP文本框的内容
        InetAddress address=InetAddress.getByName(ip);
        byte[] data=sendText.getText().getBytes();  //获取要发送的数据
        DatagramPacket dp=new DatagramPacket(data,data.length,address,8888);
                                                    //定义数据包
        String myip=InetAddress.getLocalHost().getHostAddress();  //获取本机的IP
        //将发送信息添加一信息接收文本域中
        text.append("本机-"+myip+":\n "+sendText.getText()+"\n");
        socket.send(dp);                            //发送数据包
        sendText.setText(null);
    }catch(UnknownHostException ex){
        ex.printStackTrace();
```

```
        }catch(IOException ex){
            ex.printStackTrace();
        }
    }

    @SuppressWarnings("unused")
    public static void main(String[] args) {
        JFrame frame=new UDPChat();              //生成窗体并添加窗体实例名
    }
}
```

3）运行程序

在一台计算机上运行一个程序，再在另一台计算机上运行这个程序，然后将两个程序界面"IP 文本框"中的地址信息修改为对方的 IP 地址信息，在"发送信息"文本框中，输入聊天内容，然后回车或单击"发送"按钮，都可以进行聊天。

5．任务拓展

1）解决"发送信息"框中信息为空时的发送问题

本程序中还存在一个问题，即"发送信息"框中不输入内容时，也可以发送信息，如何解决这个问题？

提示：修改 sendMsg()方法如下：

```
private void sendMsg(){
    //判断发送信息文本框中的内容是否为空
    if(!sendText.getText().equals("")){
        try{
            String ip=ipText.getText();      //获取 IP 文本框的内容
            ...                               //其他的程序行
        }
    }
}
```

2）解决"发送信息"框中信息为"纯空格"时的发送问题

如果在"发送信息"框中输入纯空格，这时还可以发送信息，但接收信息框中却看不到消息，如何解决这个问题？

提示：修改 sendMsg()方法如下：

```
private void sendMsg(){
    //判断发送信息文本框中的内容是否为空
    if(!(sendText.getText().trim()).equals("")){
        try{
            String ip=ipText.getText();      //获取 IP 文本框的内容
            ...                               //其他的程序行
        }
    }
}
```

第 15 章 多 线 程

15.1 实验目的

(1) 掌握线程创建的两种方式。
(2) 掌握线程控制的基本方法。
(3) 掌握实现线程同步的两种方式。
(4) 能够灵活运用等待和通知实现类似生产者-消费者问题。
(5) 理解同步引发的死锁问题。

15.2 实验任务

(1) 任务1：使用 Thread 和 Runnable 模拟时钟线程。
(2) 任务2：线程控制的基本方法。
(3) 任务3：模拟夫妻二人去银行取钱。
(4) 任务4：生产者-消费者问题。

15.3 实验内容

15.3.1 任务1 使用 Thread 和 Runnable 模拟时钟线程

1. 任务目的
(1) 掌握如何使用 Thread 类创建线程。
(2) 掌握如何使用 Runnable 创建线程。

2. 任务描述
通过 Thread 和 Runnable 两种方式创建时钟线程，实现每隔 1s 打印一下系统当前时间。

3. 实施步骤
1) 通过继承 Thread 类的方式实现时钟线程
step1：定义 ClockThread 类继承 Thread 类。
step2：重写 run()方法。
step3：创建 ClockThread 对象，启动线程。
参考代码如下：

```
import java.util.Date;
class ClockThread extends Thread {
    @Override
```

```java
    public void run() {
        //线程一直运行
        while (true) {
            //打印系统时间
            System.out.println(new Date());
            try {
                //线程休眠1000ms
                Thread.sleep(1000);
            } catch (InterruptedException e) {
                //注：线程休眠期间被中断时抛出此异常
                e.printStackTrace();
            }
        }
    }
}
public class Task1_1 {
    public static void main(String[] args) {
        //创建时钟线程
        ClockThread clock=new ClockThread();
        //启动线程
        clock.start();
    }
}
```

2）运行程序

观察结果，如图 15-1 所示。

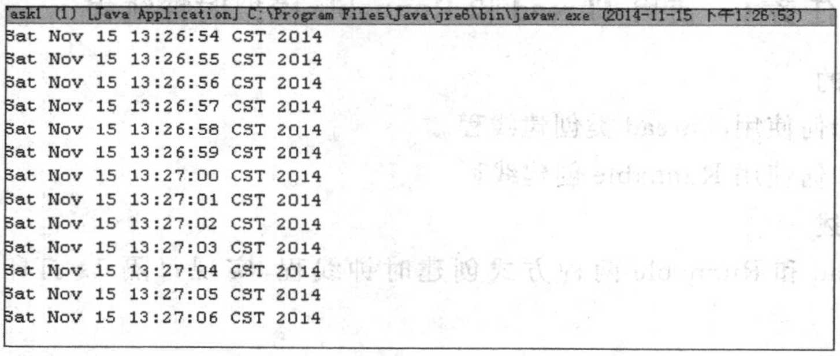

```
askl (1) [Java Application] C:\Program Files\Java\jre6\bin\javaw.exe (2014-11-15 下午1:26:53)
Sat Nov 15 13:26:54 CST 2014
Sat Nov 15 13:26:55 CST 2014
Sat Nov 15 13:26:56 CST 2014
Sat Nov 15 13:26:57 CST 2014
Sat Nov 15 13:26:58 CST 2014
Sat Nov 15 13:26:59 CST 2014
Sat Nov 15 13:27:00 CST 2014
Sat Nov 15 13:27:01 CST 2014
Sat Nov 15 13:27:02 CST 2014
Sat Nov 15 13:27:03 CST 2014
Sat Nov 15 13:27:04 CST 2014
Sat Nov 15 13:27:05 CST 2014
Sat Nov 15 13:27:06 CST 2014
```

图 15-1 时钟线程运行效果图

3）通过实现 Runnable 接口的方式实现时钟线程

step1：定义 ClockRunnable 类实现 Runnable 接口。

step2：重写 run()方法。

step3：使用接收 Runnable 接口作为参数的构造方法创建 Thread 对象，启动线程。

参考代码如下：

```java
import java.util.Date;
class ClockRunnable implements Runnable{
```

```
        @Override
        public void run() {
            //线程一直运行
            while (true) {
                //打印系统时间
                System.out.println(new Date());
                try {
                    //线程休眠 1000ms
                    Thread.sleep(1000);
                } catch (InterruptedException e) {
                    //注：线程休眠期间被中断时抛出此异常
                    e.printStackTrace();
                }
            }
        }
}
public class Task1_2 {
    public static void main(String[] args) {
        //创建时钟线程
        Thread clock=new Thread(new ClockRunnable());
        //启动线程
        clock.start();
    }
}
```

4) 讨论

为何启动线程调用 start() 方法，直接调用 run() 方法行吗？两者的区别是什么？

4. 任务拓展

(1) 同时创建两个时钟线程，交替轮流执行。

(2) 改变默认线程的名称并输出。

15.3.2　任务2　线程控制的基本方法

1. 任务目的

掌握线程控制的基本方法。

2. 任务描述

通过阅读程序，掌握线程控制的基本方法：线程休眠、线程中断和线程联合等。

3. 实施步骤

1) 线程休眠

阅读如下程序代码：

```
public class Task2_1 {
    public static void main(String[] args) {
        //创建两个线程
        Thread t1=new SleepingThread("线程 1");
```

```java
        Thread t2=new SleepingThread("线程 2");
        //启动线程
        t1.start();
        t2.start();
        //主线程休眠 1s
        try {
            System.out.println("主线程休眠 1s…");
            Thread.sleep(1000);
            System.out.println("主线程休眠结束。");
        } catch (InterruptedException e) {
            e.printStackTrace();
        }
    }
}
//演示线程休眠
class SleepingThread extends Thread {
    public SleepingThread(String name) {
        super(name);
    }
    @Override
    public void run() {
        try {
            System.out.println(Thread.currentThread().getName()+" 睡着了…");
            sleep(3000);            //线程休眠
            System.out.println(Thread.currentThread().getName()+" 睡醒了。");
        } catch (InterruptedException e) {
            //注：休眠过程中被中断时抛出此异常
            e.printStackTrace();
            System.out.println(Thread.currentThread().getName()+" 被中断了。");
        }
    }
}
```

运行程序，观察结果，如图 15-2 所示。

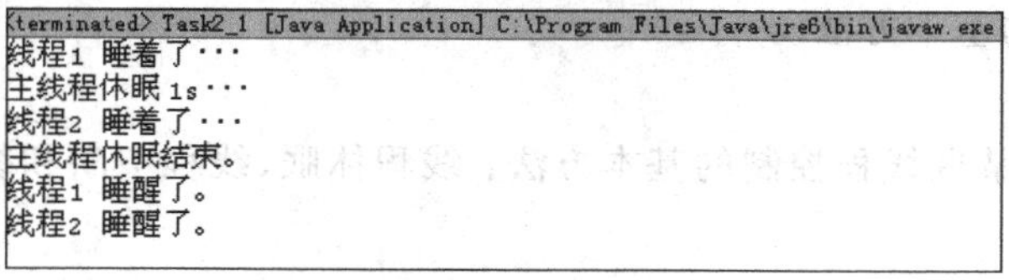

```
<terminated> Task2_1 [Java Application] C:\Program Files\Java\jre6\bin\javaw.exe
线程1 睡着了…
主线程休眠 1s…
线程2 睡着了…
主线程休眠结束。
线程1 睡醒了。
线程2 睡醒了。
```

图 15-2　演示线程休眠

分析出现上述结果 CPU 是如何调度各个线程的。

2）线程中断

阅读如下程序代码，分析其完成的功能。

```java
import java.util.Scanner;
public class Task2_2 {
    public static void main(String[] args) {
        InputMonitor inputMonitor=new InputMonitor();
        inputMonitor.start();                          //启动监控线程
        Scanner scanner=new Scanner(System.in);
        String choice="";
        if(!choice.equals("stop")){
            choice=scanner.next();
        }
        inputMonitor.interrupt();                      //如果是stop,中断线程运行
    }
}
//监控用户输入的线程
class InputMonitor extends Thread{
    private int count=1;                               //监控次数
    @Override
    public void run() {
        super.run();
        while(!isInterrupted()){
            System.out.println(getName()+"Monitoring…"+count++);
            try {
                Thread.sleep(3000);                    //休眠2s
            } catch (InterruptedException e) {
                break;                                 //停止执行
            }
        }
        System.out.println("Monitoring is stopped by user");
    }
}
```

3）线程的优先级

阅读如下程序代码：

```java
public class Task2_3 {
    public static void main(String[] args) {
        //创建线程
        Thread minPriority=new Thread(new Printer(),"最低优先级线程");
        Thread norPriority=new Thread(new Printer(),"正常优先级线程");
        Thread maxPriority=new Thread(new Printer(),"最高优先级线程");
        //设置不同的优先级
        minPriority.setPriority(Thread.MIN_PRIORITY);
        norPriority.setPriority(Thread.NORM_PRIORITY);
        maxPriority.setPriority(Thread.MAX_PRIORITY);
        //执行线程
        minPriority.start();
```

```
        norPriority.start();
        maxPriority.start();
    }
}
//打印数字
class Printer implements Runnable {
    @Override
    public void run() {
        for(int i=0; i<10; i++) {
            System.out.println(Thread.currentThread().getName()+" 打印 "+i);
        }
        System.out.println(Thread.currentThread().getName()+" 运行结束");
    }
}
```

运行程序,运行结果如图 15-3 所示。

```
<terminated> Task2_3 [Java Application] C:\Program Files\Java\jre6\bin
正常优先级线程 打印 0
正常优先级线程 打印 1
正常优先级线程 打印 2
正常优先级线程 打印 3
正常优先级线程 打印 4
正常优先级线程 打印 5
正常优先级线程 打印 6
正常优先级线程 打印 7
正常优先级线程 打印 8
正常优先级线程 打印 9
正常优先级线程 运行结束
最低优先级线程 打印 0
最高优先级线程 打印 0
最高优先级线程 打印 1
最高优先级线程 打印 2
最高优先级线程 打印 3
最高优先级线程 打印 4
最高优先级线程 打印 5
最高优先级线程 打印 6
最高优先级线程 打印 7
最高优先级线程 打印 8
最高优先级线程 打印 9
最高优先级线程 运行结束
最低优先级线程 打印 1
最低优先级线程 打印 2
最低优先级线程 打印 3
最低优先级线程 打印 4
最低优先级线程 打印 5
最低优先级线程 打印 6
最低优先级线程 打印 7
最低优先级线程 打印 8
最低优先级线程 打印 9
最低优先级线程 运行结束
```

图 15-3 演示线程的优先级

分析程序的运行结果,注意上述运行结果在不同的机器上可能不同,同时结果也说明最高优先级的并不一定就一定先运行。因此,在实际编程时,不提倡使用线程的优先级来保证算法的正确执行。

4. 任务拓展

(1) 阅读关于线程联合的代码,理解线程联合的含义。

(2) 阅读关于守护线程的代码,理解线程守护的含义。

15.3.3　任务 3　模拟夫妻二人去银行取钱

1. 任务目的

（1）理解线程同步的含义。

（2）掌握实现线程同步的两种方法。

2. 任务描述

一对夫妻在银行存了 3000 块钱，男的拿着存折，女的拿着卡。有一天，两个人同时去取钱，结果在两个人的密切配合下，竟然取出了 4000 块，账户上变成了－1000 块！你一定迫不及待地想知道他们是如何做到的吧？赶快研究一下下面的程序吧！

3. 实施步骤

（1）阅读如下程序代码：

```java
/*模拟夫妻二人提款的操作,未使用同步*/
public class Task3_1 {
    public static void main(String[] args) {
        BankAccount bankAccount=new BankAccount(3000);
        //要被执行的任务
        WithDraw withdraw=new WithDraw(bankAccount);
        Thread husband=new Thread(withdraw);
        Thread wife=new Thread(withdraw);
        husband.setName("HUSBAND");
        wife.setName("WIFE");
        husband.start();
        wife.start();
    }
}
class BankAccount{
    private int balance;
    /**
     * 新开账户
     * @param balance 初始金额
     */
    public BankAccount(int balance){
        this.balance=balance;
    }
    //查询余额
    public int getBalance(){
        return balance;
    }
    //取款
    public void withdraw(int amount){
        balance=balance-amount;
    }
```

```java
}
//代表夫妇两人都有的行为——检查余额,然后花掉,中间都会偷偷地睡一觉
class WithDraw implements Runnable{
    private BankAccount account;
    public WithDraw(BankAccount account){
        this.account=account;
    }
    @Override
    public void run() {
        if(account.getBalance()<2000){
            System.out.println(Thread.currentThread().getName()+"在查询时,钱透支了,不够了");
        }
        else
        {
            System.out.println(Thread.currentThread().getName()+"在查询时,余额为"+account.getBalance());
            System.out.println(Thread.currentThread().getName()+"准备从账户上取走2000");
            try {
                System.out.println(Thread.currentThread().getName()+"准备休息");
                Thread.sleep(500);
            } catch (InterruptedException e) {
                e.printStackTrace();
            }
            System.out.println(Thread.currentThread().getName()+"醒来");
            account.withdraw(2000);
            System.out.println(Thread.currentThread().getName()+"完成提款");
            System.out.println("现在账户上还剩"+account.getBalance()+"元");
        }
    }
}
```

(2) 运行程序,观察结果,如图 15-4 所示,分析出现问题的原因。

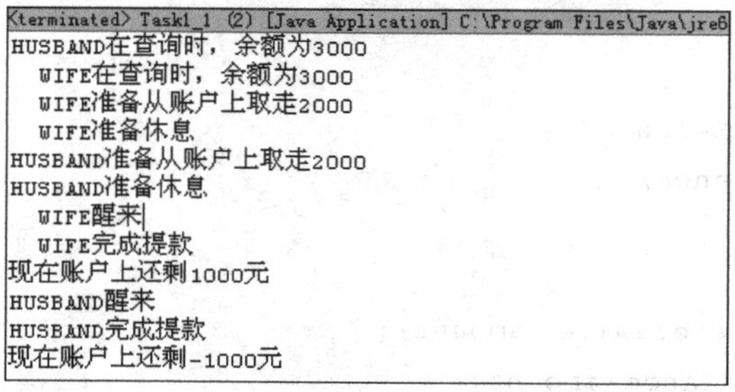

```
<terminated> Task1_1 (2) [Java Application] C:\Program Files\Java\jre6
HUSBAND在查询时, 余额为3000
    WIFE在查询时, 余额为3000
    WIFE准备从账户上取走2000
    WIFE准备休息
HUSBAND准备从账户上取走2000
HUSBAND准备休息
    WIFE醒来
    WIFE完成提款
现在账户上还剩1000元
HUSBAND醒来
HUSBAND完成提款
现在账户上还剩-1000元
```

图 15-4 未使用线程同步夫妻二人取钱的运行结果

(3) 使用同步代码块改写程序。

参考代码如下：

```java
class WithDraw implements Runnable {
    private BankAccount account;
    public WithDraw(BankAccount account) {
        this.account=account;
    }
    @Override
    public void run() {
        //对账户进行同步
        synchronized (account) {
            System.out.println(Thread.currentThread().getName()+"得到了账户的控制权——开始");
            if (account.getBalance()<2000) {
                System.out.println(Thread.currentThread().getName()
                    +"在查询时,钱透支了,不够了");
            } else {
                System.out.println(Thread.currentThread().getName()+"在查询时,余额为"+account.getBalance());
                System.out.println(Thread.currentThread().getName()
                    +"准备从账户上取走 2000");
                try {
                    System.out.println(Thread.currentThread().getName()
                        +"准备休息");
                    Thread.sleep(500);
                } catch (InterruptedException e) {
                    e.printStackTrace();
                }
                System.out.println(Thread.currentThread().getName()+"醒来");
                account.withdraw(2000);
                System.out.println(Thread.currentThread().getName()+"完成提款");
                System.out.println("现在账户上还剩"+account.getBalance()+"元");
            }
            System.out.println(Thread.currentThread().getName()+"释放了账户的控制权——结束");
        }
    }
}
```

运行程序,观察结果,如图 15-5 所示。

4. 任务拓展

使用同步方法改写程序。

15.3.4 任务 4 生产者-消费者问题

1. 任务目的

(1) 理解等待和通知的含义。

```
<terminated> Task3_2 [Java Application] C:\Program Files\Java\jre6\bin\j
HUSBAND得到了账户的控制权——开始
HUSBAND在查询时,余额为3000
最后银行账户还剩: 3000
HUSBAND准备从账户上取走2000
HUSBAND准备休息
HUSBAND醒来
HUSBAND完成提款
现在账户上还剩1000元
HUSBAND释放了账户的控制权——结束
        WIFE得到了账户的控制权——开始
        WIFE在查询时,钱透支了,不够了
        WIFE释放了账户的控制权——结束
```

图 15-5 使用同步代码块夫妻二人取钱的运行结果

(2) 掌握生产者-消费者这类问题的解决方法。

2. 任务描述

生产者 Producer 和消费者 Consumer 共享一个缓冲区 Buffer。生产者向缓冲区中写入数据,消费者从缓冲区中取出数据并进行求和运算。

3. 实施步骤

(1) 阅读如下程序代码。

先来看 Buffer 类的实现:

```java
public class Buffer {
    private int value=-1;          //缓冲区中的数据
    /**
     * 向缓冲区中写入数据。
     * @param value 待写入数据
     */
    public void setValue(int value) {
        //写入数据
System.out.println(Thread.currentThread().getName()+" 写入数据 "+value);
        this.value=value;
    }
    /**
     * 从缓冲区中取出数据。
     * @return 取出的数据
     */
    public int getValue() {
        //取出数据
System.out.println(Thread.currentThread().getName()+" 取出数据 "+value);
        return value;
    }
}
```

再来看 Producer 类和 Consumer 类的实现:

```java
public class Producer implements Runnable {
    //生产者写入数据的缓冲区
```

```java
    private final Buffer buffer;
    public Producer(Buffer buffer) {
        this.buffer=buffer;
    }
    @Override
    public void run() {
        //依次向缓冲区中写入 1~10
        for(int i=1; i<=10; i++)
            buffer.setValue(i);
    }
}
public class Consumer implements Runnable {
    //用以取出数据的缓冲区
    private final Buffer buffer;
    private int sum;
    public Consumer(Buffer buffer) {
        this.buffer=buffer;
    }
    @Override
    public void run() {
        sum=0;
        for(int i=1; i<=10; i++) {
            sum+=buffer.getValue();
            System.out.println(Thread.currentThread().getName()+" sum="+sum);
        }
    }
}
```

最后一起来看一下测试驱动类的实现:

```java
public class ProducerCunsumerNoSynNoWait {
    public static void main(String[] args) {
        //建立一个缓冲区
        Buffer buffer=new Buffer();
        //新建生产者和消费者线程,并让两者共享该缓冲区
        Thread producer=new Thread(new Producer(buffer),"生产者");
        Thread consumer=new Thread(new Consumer(buffer),"消费者");
        //启动生产者和消费者线程
        producer.start();
        consumer.start();
        //结果:出现错误
    }
}
```

(2) 运行程序,观察结果,如图 15-6 所示,分析出现问题的原因。

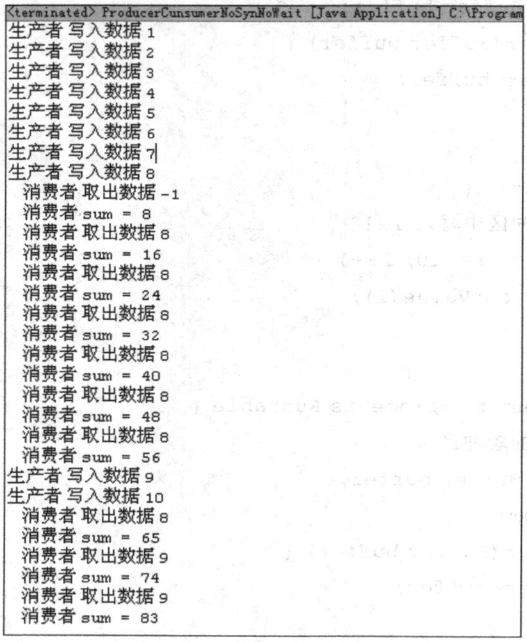

```
<terminated> ProducerCunsumerNoSynNoWait [Java Application] C:\Program
生产者 写入数据 1
生产者 写入数据 2
生产者 写入数据 3
生产者 写入数据 4
生产者 写入数据 5
生产者 写入数据 6
生产者 写入数据 7
生产者 写入数据 8
  消费者 取出数据 -1
  消费者 sum = 8
  消费者 取出数据 8
  消费者 sum = 16
  消费者 取出数据 8
  消费者 sum = 24
  消费者 取出数据 8
  消费者 sum = 32
  消费者 取出数据 8
  消费者 sum = 40
  消费者 取出数据 8
  消费者 sum = 48
  消费者 取出数据 8
  消费者 sum = 56
生产者 写入数据 9
生产者 写入数据 10
  消费者 取出数据 8
  消费者 sum = 65
  消费者 取出数据 9
  消费者 sum = 74
  消费者 取出数据 9
  消费者 sum = 83
```

图 15-6 未做任何处理的生产者-消费者问题

（3）将 Buffer 类中的方法变为同步方法。

参考代码如下：

```java
public class Buffer {
    private int value=-1;          //缓冲区中的数据
    /**
     * 向缓冲区中写入数据(同步方法)。
     * @param value 待写入数据
     */
    public synchronized void setValue(int value) {
        //写入数据
        System.out.println(Thread.currentThread().getName()+" 写入数据 "+value);
        this.value=value;
    }
    /**
     * 从缓冲区中取出数据(同步方法)。
     * @return 取出的数据
     */
    public synchronized int getValue() {
        //取出数据
System.out.println(Thread.currentThread().getName()+" 取出数据 "+value);
        return value;
    }
}
```

分析：如果仅仅只是将 Buffer 类中的方法变为同步方法，程序会不会正常运行？
(4) 加入 wait()方法和 notify()方法。

参考代码如下：

```java
public class Buffer {
    private int value=-1;                          //缓冲区中的数据
    private boolean occupied=false;                //标志缓冲区中是否有数据
    /**
     * 向缓冲区中写入数据。如果缓冲区被占用,则等待缓冲区取出数据通知,收到通知后再试图写入,
     * 写入数据后发出写入数据通知。
     * @param value 待写入数据
     * @throws InterruptedException 线程等待通知时被中断抛出此异常。
     */
    public synchronized void setValue(int value) throws InterruptedException {
        //准备写入数据
        System.out.println(Thread.currentThread().getName()+" 准备写入数据…");
        while(occupied) {
            //当缓冲区被占用时,当前线程暂时释放缓冲区的锁,等待数据取出通知
            System.out.println(Thread.currentThread().getName()+" 等待写入数据");
            wait();
        }
        //收到取出数据通知后,发现缓冲区为空,继续写入数据
        System.out.println(Thread.currentThread().getName()+" 写入数据 "+value);
        this.value=value;
        //设置缓冲区中有数据的标志
        occupied=true;
        //发出写入数据通知,通知等待取出数据的线程
        notifyAll();
    }
    /**
     * 从缓冲区中取出数据。如果缓冲区已空,则等待写入数据通知,收到通知后再试图取出数据,
     * 取出数据后发出缓冲区取出数据通知。
     * @return 取出的数据
     * @throws InterruptedException 线程等待通知时被中断抛出此异常。
     */
    public synchronized int getValue() throws InterruptedException {
        //准备取出数据
        System.out.println(Thread.currentThread().getName()+" 准备取出数据…");
        while(!occupied) {
            //当缓冲区为空时,当前线程暂时释放缓冲区的锁,等待数据写入通知
            System.out.println(Thread.currentThread().getName()+" 等待取出数据");
            wait();
        }
        //收到写入数据通知后,发现缓冲区中有数据,继续取出数据
        System.out.println(Thread.currentThread().getName()+" 取出数据 "+value);
```

```
            //设置缓冲区为空的标志
            occupied=false;
            //发出取出数据通知,通知等待写入数据的线程
            notifyAll();
            //返回取出的数据
            return value;
        }
    }
```

(5) 运行程序,观察结果。

(6) 讨论:如果把其中的 while 换成 if,会怎样?观察输出结果并分析。

4. 程序拓展

(1) 继续修改程序,引入多个生产者和消费者,观察程序的运行结果。

(2) 修改课本中的猜数字的例子,使之每次运行都能得出正确的运行结果。

第 16 章　数据库操作

16.1　实验目的

(1) 掌握 JDBC 的工作原理。
(2) 掌握 JDBC 数据库常用数据连接方式。
(3) 掌握 JDBC 中与数据操作相关的类和接口的常用方法。
(4) 掌握灵活使用 SQL 语句实现数据的增、删、改、查功能。
(5) 掌握预处理对象和存储过程的使用方法。
(6) 掌握使用集合处理数据库查询结果的方法。

16.2　实验任务

任务：电子信息协会会员管理信息系统 AMIS。

16.3　实验内容

任务　电子信息协会会员管理信息系统 AMIS

1. 任务目的

(1) 掌握使用纯 JDBC 驱动直接连接 MySQL 数据库的方法。
(2) 掌握 JDBC 中与数据操作相关的类和接口的常用方法。
(3) 掌握使用 SQL 语句实现数据的增、删、改、查功能。
(4) 掌握预处理对象的使用方法。
(5) 掌握使用集合处理数据库查询结果的方法。

2. 任务描述

某高校的电子信息协会现在还在用纸质介质对该协会的会员信息进行管理，现在在学校的大力支持下，该协会具有了一台自己的服务器，经过全会人员投票决定，要开发一套会员管理信息系统(Academician Management Information System，AMIS)。现阶段该系统只要能简单记录每个会员的基本信息，在需要时能方便地查询每个会员的信息即可。

3. 任务分析与设计

根据任务要求，可知这个系统要对会员 Academician 进行信息管理，要能录入会员信息(添加)，能修改会员信息(修改)，能查看所有的会员信息(查询)，能删除过期的数据(删除)(注意实际的应用实例中，数据非常重要，最好不要删除数据，设置为删除状态即可)。

因此设计会员 Academician 的数据结构如表 16-1 所示。

表 16-1　会员 Academician 的表结构

序号	字段	类型(长度)	键值	可空	说　明
1	id	int(4)	pk	not	微机编码,自动增长,主键
2	stuName	varchar(20)		not	会员姓名
3	stuDept	varchar(20)			会员所在系别
4	stuClass	varchar(20)			会员所在班级
5	inDt	datetime			入会时间
6	outDt	datetime			出会时间
7	bLeave	bit(1)			是否在会,0 表示在会中,1 表示离会,默认为 0
8	tell	varchar(20)			会员的电话

因为 MySQL 数据库比较小巧,对于个人用户来说也完全免费,因此本任务所采用的数据库为 MySQL 5.5,数据库的前端管理工具使用 Navicat 10.0 for MySQL,在 MySQL 中创建 amis 数据库。

4. 实施步骤

1) 创建 amis 数据库

在使用 MySQL 5.5 前要安装 MySQL 5.5(安装版的下载地址为 http://dev.mysql.com/downloads/windows/installer/)和 Navicat 10.0,这个工作大家要提前完成。打开 Navicat 10.0 平台,默认没有任何连接,如图 16-1 所示。

图 16-1　Navicat 10.0 的管理界面

在 Navicat 左侧窗口右击,打开如图 16-2 所示的菜单,选择"新建连接",打开如图 16-3 所示的"新建连接"对话框,填写连接名和密码后,单击"连接测试"按钮,弹出"连接成功"对话框。

单击"确定"按钮后,创建一个与 MySQL 的连接实例 local,双击 local,打开该连接,可以看到该连接下的数据库列表如图 16-4 所示。

在 local 根目录上右击,弹出如图 16-4 所示的右键菜单,单击"新建数据库"菜单,打开如图 16-5 所示的"新建数据库"对话框,输入数据库名 amis,选择字符集 utf8 或是 gbk,使数据库能支持中文,设置完成后,单击"确定"按钮,则会添加 amis 数据库。

注意:在创建数据库时,一定要注意选择字符集为支持中文的字符集 utf8 或是 gbk,否则数据库在创建完成后,可能会因未对 MySQL 的默认字符集做过配置而不支持中文。

图 16-2 新建连接菜单

图 16-3 "新建连接"对话框

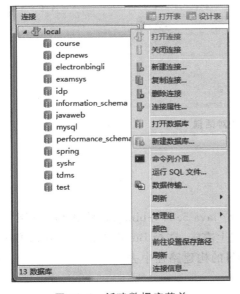

图 16-4 新建数据库菜单

图 16-5 "新建数据库"对话框

2）创建 Academician 数据表

在上一步创建的完成的数据库 amis 上双击，打开该数据库，在表上右击，打开"新建表"菜单，如图 16-6 所示。

在图 16-6 中单击"新建表"菜单，打开图 16-7 所示的"新建表"对话框，按表 16-1 中的字

· 165 ·

段内容和图 16-7 的格式设置各个字段，设置完成后，单击工具栏上的"保存"按钮，打开保存数据表对话框，在表名中输入 Academician 后，单击"确定"按钮，如图 16-8 所示，单击"确定"按钮，即可以完成数据表的创建。

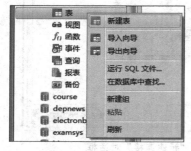

图 16-6　新建表菜单

图 16-7　新建表对话框

图 16-8　保存数据表对话框

3）创建 Java 项目

在 Eclipse 中创建 Java 项目 Lab16。

4）添加 JDBC 驱动包

本任务用到的 JDBC 驱动包是 mysql-connector-java-5.1.7-bin.jar，将该包"复制"→"粘贴"到 Eclipse 中的 Lab16 项目中去，然后在该包上右击，选择 Build Path→Add to Build Path 菜单，单击后，Eclipse 会将该驱动包添加到项目的构建路径中去。

5）创建包

在项目 Lab16 中创建包 task。

6）创建数据库连接操作父类 DBConnection

在包 task 中创建数据文件包 dao，在 dao 包中创建文件 DBConnection.java，负责打开数据库连接与关闭各个数据操作对象，注意处理异常，修改该文件的内容如下：

```
package task;
```

图 16-9　添加驱动包并将其添加到构建路径中去

```java
import java.sql.Connection;
import java.sql.DriverManager;
import java.sql.PreparedStatement;
import java.sql.ResultSet;
import java.sql.Statement;
/**
 * 数据库连接操作父类,负责打开数据连接,关闭数据库操作对象
 * @author sf
 */
public class DBConnection {
    //连接属性定义区
    private final static String CLS="com.mysql.jdbc.Driver";          //驱动包名
    private final static String URL="jdbc:mysql://localhost:3306/amis";  //URL 名称
    private final static String USER="root";                          //数据库访问用户名
    private final static String PWD="123456";                         //数据库访问密码

    //公共数据库操作对象
    public static Connection conn=null;                  //连接对象
    public static Statement stmt=null;                   //命令集对象
    public static PreparedStatement pStmt=null;          //预编译命令集对象
    public static ResultSet rs=null;                     //结果集对象

    /**
     * 打开连接的方法
     */
    public static void getConnection(){
        try{
            Class.forName(CLS);                                          //加载驱动类
            conn=DriverManager.getConnection(URL,USER,PWD);              //打开连接
        }catch(Exception ex){
            ex.printStackTrace();
        }
    }

    /**
     * 关闭所有的数据库操作对象的方法
     */
```

```
public static void closeAll(){
    try{
        if(rs!=null){                                      //关闭结果集
            rs.close();
            rs=null;
        }
        if(stmt!=null){                                    //关闭命令集
            stmt.close();
            stmt=null;
        }
        if(pStmt!=null){                                   //关闭预编译命令集
            pStmt.close();
            pStmt=null;
        }
        if(conn!=null){                                    //关闭连接
            conn.close();
            conn=null;
        }
    }catch(Exception ex){
        ex.printStackTrace();
    }
}
```

7）创建实体类

在 task 包中创建 entity 包，在 entity 包中创建会员信息实体类文件 Academician.java，参考数据表定义私有属性，使用环境生成 getter 和 setter 方法，修改该类内容如下：

```
package task.entity;
/**
 * 会员信息实体类
 * @author sf
 */
public class Academician {
    private int id;                        //微机编码,自动增长
    private String stuName;                //会员名
    private String stuDept;                //会员所在系别
    private String stuClass;               //会员所在班级
    private String inDt;                   //入会时间
    private String outDt;                  //离会时间
    private boolean bLeave;                //是否离会
    private String tell;                   //会员联系电话
    public Academician() {                 //无参构造方法
    }
    /**带参构造方法,在生成实例时对数据进行初始化*/
    public Academician(int id,String stuName,String stuDept,
        String stuClass,String inDt,String outDt,boolean bLeave,
```

```java
            String tell) {
        this.id=id;
        this.stuName=stuName;
        this.stuDept=stuDept;
        this.stuClass=stuClass;
        this.inDt=inDt;
        this.outDt=outDt;
        this.bLeave=bLeave;
        this.tell=tell;
    }

    public int getId() {
        return id;
    }

    public void setId(int id) {
        this.id=id;
    }

    public String getStuName() {
        return stuName;
    }

    public void setStuName(String stuName) {
        this.stuName=stuName;
    }
    //使用Eclipse环境生成getter和setter方法即可,这里省略其他的getter和setter方法
}
```

注意：在eclipse中使用属性生成getter和setter方法的过程如下：在类文件编辑界面中右击→选择Source→选择Generate Getters and Setters菜单，如图16-10所示。

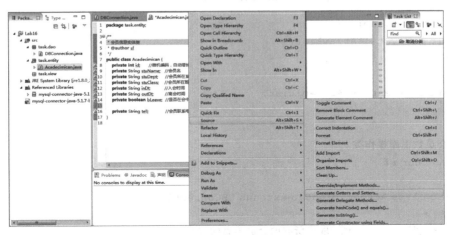

图16-10 选择Generate Getters and Setters菜单

单击菜单后,打开生成 Generate Getters and Setters 对话框,如图 16-11 所示。

图 16-11　在 Eclipse 中生成 getters 和 setters 方法

单击 Sellect All 按钮,再单击 OK 按钮,即可以生成实体类的 getters 和 setters 方法。

8) 创建数据库操作类

在 dao 包中创建数继承自数据库连接操作类 DBConnection 的会员数据库操作类文件 AcademicianDao.java,在类文件 AcademicianDao.java 添加 CRUD(插入数据、读取数据、更新数据、删除数据)几个方法,在读取数据时,使用 List 集合框架处理从数据库读取的结果,修改后类文件 AcademicianDao.java 的内容如下:

```
package task.dao;
import java.util.ArrayList;
import java.util.List;

import task.entity.Academician;

/**
 * 会员数据库操作类
 * @author sf
 */
public class AcademicianDao extends DBConnection {
    /**
     * 使用命令集实现:取得所有的会员信息的方法
     * @return
     */
```

```java
public List<Academician>getAllAcademicianList(){
    List<Academician>list=new ArrayList<Academician>();
    try{
        getConnection();                                //打开数据库连接
        stmt=conn.createStatement();                    //命令集
        String sql="select * from Academician";         //查询字符串
        rs=stmt.executeQuery(sql);                      //执行命令返回结果集
        while(rs.next()){                               //在循环中对结果集进行处理
            Academician item=new Academician();
            item.setId(rs.getInt("id"));
            item.setStuName(rs.getString("stuName"));
            item.setStuDept(rs.getString("stuDept"));
            item.setStuClass(rs.getString("stuClass"));
            item.setInDt(rs.getString("inDt"));
            item.setOutDt(rs.getString("outDt"));
            item.setbLeave(rs.getBoolean("bleave"));
            item.setTell(rs.getString("tell"));

            list.add(item);                             //将对象添加到集合中去
        }
    }catch(Exception ex){
        ex.printStackTrace();
    }finally{
        closeAll();                                     //关闭所有的数据库操作对象
    }
    return list;
}

/**
 * 添加会员信息的方法
 * @param item 要添加的会员信息对象
 * @return int 影响的行数
 */
public int addAcademician(Academician item){
    int iRow=0;
    try{
        getConnection();                                //打开数据库连接
        //查询语句
        String sql="insert into Academician(stuName,stuDept,stuClass,inDt,outDt,bLeave,tell) values(?,?,?,?,?,?,?)";
        pStmt=conn.prepareStatement(sql);               //预编译命令集
        pStmt.setString(1,item.getStuName());
        pStmt.setString(2,item.getStuDept());
        pStmt.setString(3,item.getStuClass());
        pStmt.setString(4,item.getInDt());
```

```java
            pStmt.setString(5,item.getOutDt());
            pStmt.setBoolean(6,item.isbLeave());
            pStmt.setString(7,item.getTell());

            iRow=pStmt.executeUpdate();                    //更新数据库
        }catch(Exception ex){
            ex.printStackTrace();
        }finally{
            closeAll();                                    //关闭所有的数据库操作对象
        }
        return iRow;
    }

    /**
     * 修改会员信息的方法
     * @param item 要修改的会员信息对象
     * @return int 影响的行数
     */
    public int editAcademician(Academician item){
        int iRow=0;
        try{
            getConnection();                               //打开数据库连接
            String sql="update Academician set stuName=?,stuDept=?,"
                    +"stuClass=?,inDt=?,outDt=?,"+"bLeave=?,"
                    +"tell=? where id=?";                  //查询语句
            pStmt=conn.prepareStatement(sql);              //预编译命令集
            pStmt.setString(1,item.getStuName());
            pStmt.setString(2,item.getStuDept());
            pStmt.setString(3,item.getStuClass());
            pStmt.setString(4,item.getInDt());
            pStmt.setString(5,item.getOutDt());
            pStmt.setBoolean(6,item.isbLeave());
            pStmt.setString(7,item.getTell());
            pStmt.setInt(8,item.getId());

            iRow=pStmt.executeUpdate();                    //更新数据库
        }catch(Exception ex){
            ex.printStackTrace();
        }finally{
            closeAll();                                    //关闭所有的数据库操作对象
        }
        return iRow;
    }

    /**
```

```java
 * 删除会员信息的方法
 * @param id 要删除的会员信息的id
 * @return int 影响的行数
 */
public int delAcademician(int id){
    int iRow=0;
    try{
        getConnection();                                //打开数据库连接
        String sql="delete from Academician where id=?";  //查询语句
        pStmt=conn.prepareStatement(sql);               //预编译命令集
        pStmt.setInt(1,id);                             //向预编译指令集中设置参数
        iRow=pStmt.executeUpdate();                     //执行更新并取得影响行数
    }catch(Exception ex){
        ex.printStackTrace();
    }finally{
        closeAll();                                     //关闭所有的数据库操作对象
    }
    return iRow;
}
```

9) 创建会员管理类进行操作测试

在 task 中创建 view 包,在 view 包中创建会员管理类文件 AMis.java,在界面中实现显示会员信息列表、添加会员、修改会员、删除会员的操作。修改后的会员管理类文件 AMis.java 文件的内容如下:

```java
package task.view;
import java.util.List;

import task.dao.AcademicianDao;
import task.entity.Academician;

/**
 * 会员管理系统测试类
 * @author sf
 */
public class AMis {
    public static void main(String[] args) {
        AcademicianDao dao=new AcademicianDao();        //数据库操作实例
        //1.显示所有的会员信息
        System.out.println("会员列表信息如下: ");
        showAcademicianList(dao.getAllAcademicianList());
        //2.添加会员信息
        Academician item=new Academician(0,"王永","信息工程系","2013级2班",
                    "2013-10-1 00:00:00",null,false,"13564563210");
        dao.addAcademician(item);
```

```java
            item=new Academician(0,"江南","信息工程系","2013级2班","2013-10-2
                        00:00:00",null,false,"13844563210");
            dao.addAcademician(item);
            //2.1 显示添加后所有的会员信息
            System.out.println("添加会员后的会员列表信息如下：");
            showAcademicianList(dao.getAllAcademicianList());
            //3.修改指定的会员信息
            item.setId(5);                                          //设置id,用于修改
            item.setOutDt("2014-12-12 00:00:00");
            item.setbLeave(true);
            dao.editAcademician(item);
            //3.1 显示修改后所有的会员信息
            System.out.println("修改会员后的会员列表信息如下：");
            showAcademicianList(dao.getAllAcademicianList());
            //4.删除指定的会员信息
            dao.delAcademician(5);
            //4.1 显示删除后所有的会员信息
            System.out.println("删除会员后的会员列表信息如下：");
            showAcademicianList(dao.getAllAcademicianList());
        }

        /**
        * 显示会员信息列表的方法
        */
        public static void showAcademicianList(List<Academician>list){
            System.out.println("id\t会员名\t会员所在系院\t会员所在班级\t入会时间\t
                        离会时间\t是否在会\t联系电话");
            for(Academician item : list){                           //在foreach循环中显示

                System.out.println(item.getId()+"\t"+item.getStuName()+"\t"+
                            item.getStuDept()+"\t"+item.getStuClass()+
                            "\t"+item.getInDt()+"\t"+item.getOutDt()+"\t"+
                            get_str_from_bLeave(item.isbLeave())+"\t"+item.getTell());
            }
        }


        /**
        * 从是否离会状态中取得是否离会的字符串
        * @param bleave 离会状态
        * @return 离会字符串
        */
        public static String get_str_from_bLeave(boolean bleave){
            String str="在会中";
            if(bleave){
                str="离会";
```

```
        }
        return str;
    }
}
```

10）运行程序

运行 AMis 文件,将会看到如图 16-12 所示的结果。

图 16-12 AMis 程序的运行结果

11）思考

再次运行程序,将会看到什么结果?为什么会出现这样的结果?

5. 任务拓展

(1) 将当前的这个控制台系统更新为可视化 UI 界面。

(2) 现在的这个系统只是完成了会员信息的管理,请为这个系统添加协会信息管理。

(3) 请为这个系统添加系/院信息管理功能。

(4) 请为这个系统添加班级信息管理功能。

(5) 以完成的系统功能为参考,完成学生学籍信息管理功能。

(6) 以完成的系统功能为参考,完成诸如员工信息管理等 MIS 系统的功能。

第二篇 综合实例篇

本篇综合前面各章节的理论与实践知识,采用 MVC 设计模式进行分析、设计并编程实现了一个 C/S 结构、图形化界面的图书管理系统。

该综合实例介绍了系统中使用的对象、业务需求、系统功能模块、系统数据结构及数据库的分析方法。

按系统的功能模块的划分,分别介绍了用户登录模块、用户管理模块、用户密码管理模块、读者信息管理模块、图书信息管理模块、图书借阅/归还操作模块、罚款管理模块、报表打印模块、帮助管理模块、主界面管理模块的设计与实现方法。

在报表打印模块中介绍了 ireport 组件的使用方法。

在帮助管理模块中介绍了"关于界面"、"帮助文件"页面的布局、使用 EasyChm 制作帮助手册的方法。

通过该综合实例的练习,读者可以熟练地掌握面向对象的软件的分析、面向对象的软件设计,以及使用 Java 开发平台对图书管理系统的各个功能实现方法,还能掌握实际应用软件中的报表、帮助文件的生成方法,认真地完成这个综合实例,能积累丰富的开发经验。

该综合实例可以作为"Java 程序设计——课程设计类"设计课题的参考项目,供个人积累开发经验;也可以作为一个小型团队项目,供小型开发团队积累团队开发经验。

第17章 图书管理系统

17.1 图书管理系统业务需求分析

为了对学校图书馆藏书及借阅者进行统一方便的管理,开发图书管理系统。根据图书馆日常工作需要,确定系统要实现6个基本功能:图书信息管理、读者信息管理、借书/还书操作、数据查询、报表打印和罚款管理等,该系统具有强大的可靠性和方便性,达到管理者和读者的满意。在设计中,要开发出界面友好、易于操作的图书管理软件,管理好图书馆中的各类信息,使图书管理工作规范化、系统化,提高信息处理的准确性,提高为读者服务的质量。

17.1.1 系统使用对象分析

对图书管理系统进行系统调查和分析,其中主要涉及的使用人员分为以下3类。

1. 系统管理员

维护整个系统的正常运行,及时更新系统信息,管理用户角色、分配权限,建立新读者信息,删除注销读者信息,备份数据,保证网络通畅和信息可靠完整。

2. 图书管理员

维护图书信息、读者信息、出版社信息、部门信息等基础数据,完成图书的入库、出库,实现读者借阅、归还、续借图书、查阅信息等操作,处理图书罚款操作,实现图书流通管理工作。

3. 读者

利用图书管理系统可以方便查询图书信息、借阅信息和罚款信息,读者包括学生和老师,不同类型的读者借阅书籍的数量和期限各有不同,由系统管理员负责进行系统设置。

17.1.2 业务需求分析

图书管理系统利用计算机等先进的信息技术和手段,实现图书和读者信息资料管理,实现借还图书、查询信息等各个环节操作。具体实现业务过程有8个。

(1) 图书管理员为读者建立读者账户,建立读者个人信息档案,其中学生账号包括学号、姓名、班级、专业、联系电话等信息,教师账号包括教工号、姓名、性别、院系、联系电话等信息。

(2) 当学生离校后,图书馆管理员要及时删除该学生的相关信息。

(3) 图书管理员对入库新图书信息进行录入,图书基本信息有图书索引号、图书名、作者、出版社、出版日期、图书类别和使用状态。对于已有同类图书进行入库,可直接增加数量,以方便读者图书借阅。

(4) 图书管理员实现读者借书、还书的操作;借书时输入借书证号,验证证号的有效性,若有效,检查读者借阅图书数量是否超过规定允许借阅图书数量,没有超过规定借阅数量,则增加借阅信息,并更新用户借阅信息和图书馆藏书信息,完成借阅操作;还书根据图书编号,找到相应借阅者,检验图书是否借阅超期,未超期则删除借阅信息,超期则进行罚款处理。

(5) 读者可以借出、归还,利用计算机方便查询图书、借阅信息和罚款信息。
(6) 对于图书超期未还、图书损坏、图书丢失等情况,进行相应的罚款处理。
(7) 当图书被丢失或图书下架,图书管理员要及时修改相应图书信息。
(8) 系统管理员负责添加、删除、修改用户,设置用户权限。

17.1.3 系统功能模块分析

面向对象程序设计中,系统设计采用 MVC 模式,本系统采用 C/S 模式的三层结构,系统的类分为 3 种:用户界面类、业务处理类、数据访问类。在系统中登录界面、图书操作界面等用户界面属于用户界面类;负责系统中的业务逻辑处理,如图书入库、借还操作等属于业务处理类;数据库访问类则实现与数据库连接,保存处理结果类。

图书管理系统的功能结构图如图 17-1 所示。

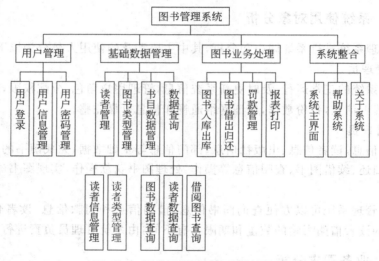

图 17-1　图书管理系统的功能结构图

17.1.4 系统数据库分析

在图书管理系统中需要将相关数据存储在不同的表中,通过不同的操作界面,发送不同的 SQL 语句命令传送到数据库表中,对数据库中对应的表进行操作,实现信息的增加、删除、修改、查询等操作,完成图书管理信息系统。

1. 分析设计数据库概念模型

数据库中概念模型常用 E-R 模型,即实体-联系模型,分析系统中主要涉及实体、实体相关属性及实体间联系。经分析,系统概要 E-R 模型如图 17-2 所示。

图 17-2　系统概要 E-R 模型

系统总体 E-R 模型如图 17-3 所示。

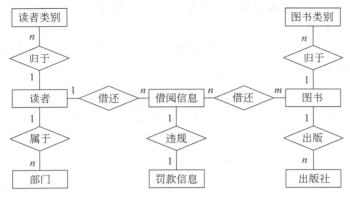

图 17-3　系统总体 E-R 模型

2．分析设计数据库逻辑模型

数据库逻辑结构设计就是将 E-R 图转换为关系模式，按照关系模式规范，将 E-R 图中的各个实体、联系转换成相应的表，也就是关系，对关系进行优化设计，符合关系范式要求。在该系统中设计关系模式如下。

(1) 用户信息表(用户编号,用户名,密码,用户类型,启用日期,状态)。

(2) 出版社(出版社编号,出版社名,出版社简称,出版社地址,联系人,联系电话,E-mail)。

(3) 部门(部门编号,部门名称,部门负责人,联系人,电话)。

(4) 图书类型(图书类型编号,图书类型名称,说明)。

(5) 图书信息(图书编号,图书名称,作者,出版社编号,ISBN,出版日期,图书页数,价格,类型代码,现存数量,简介,借阅次数)。

(6) 读者类型(读者类型编号,读者类型,借阅数量,借书期限,超期日罚金额,有效期)。

(7) 读者信息(借书证号,姓名,性别,出生日期,读者类型,证件号码,联系电话,部门编号,使用状态)。

(8) 借阅信息(借阅编号,读者借书证号,图书编号,借出日期,应还日期,罚款金额)。

(9) 罚款类型(罚款类型编号,罚款种类,罚款原因,罚款基数,罚款倍数)。

(10) 罚款信息(罚款编号,借书证号,图书编号,罚款类型,罚款金额,罚款日期)。

3．分析数据库物理结构

在数据库逻辑结构的基础上，采用 Microsoft SQL Server 2005 数据库管理系统，确定数据在物理设备上存储结构和存取方法，实现有效安全的数据管理。

17.2　功能模块实现

前面划分出图书管理系统的各个功能模块，在下面将一一进行实现。

17.2.1　用户登录模块设计

1．功能描述与分析

用户使用"用户登录"模块进入图书管理系统，在登录界面中输入用户名和密码，系统通过与数据库建立连接,查询数据库中用户表,验证用户输入信息的合法性和有效性,如果通过验证,则提示登录成功,进入主界面。如果未通过验证,则显示登录失败。

2. 涉及的数据库表

在 Microsoft SQL Server 2005 中创建图书管理数据库 bookmanage,建立用户信息表 userInfo,表结构如表 17-1 所示。

表 17-1　用户信息表结构

字段名	数据类型	长度	允许空	是否主键	含义
user_id	int	4B	不允许	是	用户编号
user_name	varchar	20 字符	不允许		用户名
user_pass	varchar	20 字符	允许		用户密码
user_type	varchar	20 字符	允许		用户类型编号
startdate	datetime	8B	允许		启用日期
valid	bit	1B	允许		是否有效

3. 界面设计分析

设计用户登录界面如图 17-4 所示,在该界面中输入用户名和密码信息,并给出相应的提示。

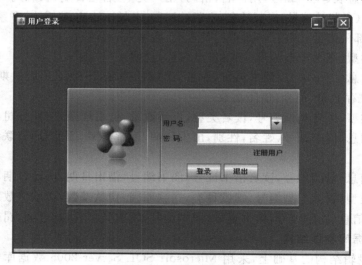

图 17-4　用户登录界面

当用户名为空或密码为空时,提示用户相应信息如图 17-5 所示。

当用户名和密码输入有误时,弹出提示框,如图 17-6 所示。

图 17-5　输入信息为空时的提示信息　　图 17-6　登录信息有误时的提示信息

4. 分析设计相关类

在用户登录界面操作中涉及的主要类及类间关系,用 UML 图描述如图 17-7 所示。

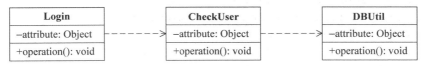

图 17-7　用户登录相关类 UML 图

1) 数据库操作类

与数据库建立连接,打开数据库连接,关闭数据库连接,实现按条件查询。该类图如图 17-8 所示。

2) 用户登录类

该类实现在数据库中用户表获取指定"用户名"和"密码"的用户数据。该类图如图 17-9 所示。

3) 用户登录界面类

设计用户登录界面类,利用各种组件和不同布局方式,设计登录界面,并实现界面上各个按钮控件事件,调用用户登录类中的方法,实现在数据库中检查该用户的合法性。该类图如图 17-10 所示。

```
┌─────────────────────────────────┐
│           DBUtil                │
├─────────────────────────────────┤
│ –driveName: String              │
│ –url: String                    │
│ –userName: String               │
│ –password: String               │
├─────────────────────────────────┤
│ +getConnection(): Connection    │
│ +closeCon(): void               │
│ +closeSm(): void                │
│ +closeRs(): void                │
└─────────────────────────────────┘
```

图 17-8　"数据库操作类"类图

```
┌─────────────────────────────────┐
│            Login                │
├─────────────────────────────────┤
│ -jbSubmit, jbClose: JButton     │
│ -img: Image                     │
│ -box, box1, box2: Box           │
│ -mainPanel: JPanel              │
│ -jtfName: JComboBox             │
│ -jpfPwd: JPasswordField         │
│ -jrb1, jrb2: JRadioButton       │
│ -bg: ButtonGroup                │
│ -zhuce: JLabel                  │
│ -jop: JOptionPane               │
│ -cu: int                        │
├─────────────────────────────────┤
│ +Login()                        │
│ +actionPerformed(): void        │
│ +mouseClicked(): void           │
│ +mouseExited(): void            │
│ +mouseEntered(): void           │
│ +mousePressed(): void           │
│ +mouseReleased(): void          │
└─────────────────────────────────┘
```

```
┌─────────────────────────┐
│       CheckUser         │
├─────────────────────────┤
│ –user_type: String      │
│ –list: List             │
├─────────────────────────┤
│ +check(): boolean       │
└─────────────────────────┘
```

图 17-9　"用户登录类"类图　　　图 17-10　"用户登录界面类"类图

5. 相关类主要代码

该模块设计涉及类的主要代码如下,详见附带代码(可从清华大学出版社网站下载)。

数据库操作 DBUtil:

```java
public class DBUtil {
    private static String driverName="com.microsoft.sqlserver.jdbc.
        SQLServerDriver";
    private static String url="jdbc:sqlserver://localhost:1434;DatabaseName=
```

```java
        bookmanage";
    private static String userName="sa";
    private static String password="123456";
        public static Connection getConnection(){
        Connection conn=null;
        try {
            Class.forName(driverName);
        } catch (ClassNotFoundException e) { e.printStackTrace();}
        try {
            conn=DriverManager.getConnection(url,userName,password);
        } catch (SQLException e) {   e.printStackTrace();  }
        return conn;
    }
        public static void closeCon(Connection con){
        if(con!=null){
            try {
                con.close();
            } catch (SQLException e) {}
        }
    }
        public static void closeSm(Statement sm){
        if(sm!=null){
            try {
                sm.close();
            } catch (SQLException e) {}
        }
    }
        public static void closeRs(ResultSet rs){
        if(rs!=null){
            try {
                rs.close();
            } catch (SQLException e) {}
        }
    }
}
```

用户登录类 CheckUser：

```java
public class CheckUser {
    public static String user_type="";
    JOptionPane jop=new JOptionPane();
    List list=new ArrayList();
    //判断用户名或密码是否正确
    public boolean check(String name, char[] cs){
        Connection con=DBUtil.getConnection();
        String pass=String.valueOf(cs);
```

```java
        try {
            String sql="select * from userInfo where user_name='"+name+"' and user_pass='"+pass+"'";
            Statement sm=con.createStatement();
            ResultSet rs=sm.executeQuery(sql);
            if(rs.next()){
                if(rs.getString("user_type")!=null){
                    user_type=rs.getString("user_type");
                }
                return true;
            }
        } catch (SQLException e) {
                e.printStackTrace();
        }finally{
            try {
                con.close();
            } catch (SQLException e) {
                e.printStackTrace();
            }
        }
        return false;
    }
}
```

用户登录界面类 Login：

```java
public class Login extends JFrame implements ActionListener,MouseListener{
    JButton jbSubmit,jbClose;
    Box box,box1,box2;
    Image img;
    JPanel mainPanel;
    JComboBox jtfName;
    JPasswordField jpfPwd;
    JRadioButton jrb1,jrb2;
    ButtonGroup bg;
    JLabel zhuce;
    JOptionPane jop=new JOptionPane();
    CheckUser cu=new CheckUser();
public Login() {
        super("用户登录");
        setLayout(null);
        ImageIcon ic=new ImageIcon("login.png");
        JLabel jl=new JLabel(ic);
        jl.setBounds(0, 0, 640, 403);
        mainPanel=(JPanel) getContentPane();
        mainPanel.setOpaque(false);
```

```java
        mainPanel.add(jl);
        jbSubmit=new JButton("登录");
        jbSubmit.addActionListener(this);
        jbClose=new JButton("退出");
        jbClose.addActionListener(this);
        box=Box.createHorizontalBox();
        box.add(jbSubmit);
        box.add(Box.createHorizontalStrut(5));
        box.add(jbClose);
        box.setBounds(303, 245, 145, 30);
        mainPanel.add(box);
        jtfName=new JComboBox(cu.readerUserName());
        jtfName.setEditable(true);
        jpfPwd=new JPasswordField();
        box2=Box.createVerticalBox();
        box2.add(jtfName);
        box2.add(Box.createVerticalStrut(5));
        box2.add(jpfPwd);
        box2.setBounds(320, 158, 150, 55);
        mainPanel.add(box2);
        zhuce=new JLabel("注册用户");
        zhuce.setBounds(420, 220, 80, 12);
        zhuce.setForeground(Color.blue);
        zhuce.addMouseListener(this);
        mainPanel.add(zhuce);
        getLayeredPane().add(jl, new Integer(Integer.MIN_VALUE));
        setBounds(400, 150, 595, 540);
        setResizable(false);
        setVisible(true);
        setDefaultCloseOperation(JFrame.EXIT_ON_CLOSE);
    }
    public void actionPerformed(ActionEvent e) {
        if(e.getSource()==jbClose){
            System.exit(0);
        }
        if(e.getSource()==jbSubmit){
            if(jtfName.getSelectedItem().equals("")||jpfPwd.getPassword()
                .length==0){
                JOptionPane.showMessageDialog(jop,"用户名和密码不能为空!");
            }else{
                boolean f;
                f=cu.check((String)jtfName.getSelectedItem(), jpfPwd.getPassword());
                if(f){
                    cu.writeUserName((String)jtfName.getSelectedItem());
                    MainInterface.status=(String)jtfName.getSelectedItem();
```

```
            setVisible(false);
            new MainInterface();
        }else{
            JOptionPane.showMessageDialog(jop, "登录失败,用户名或密码有误!");
        }
    }
    jpfPwd.setText("");
}
```

17.2.2 用户管理模块设计

1. 功能描述与分析

系统管理员使用该模块完成对"用户信息"表中的数据的增、删、改操作,即实现注册新的图书管理员/系统管理员、修改已有用户信息、删除已有用户信息等功能。

2. 涉及的数据库表

该功能模块仍使用用户信息表 userInfo。

3. 界面设计分析

使用该界面可以对图书管理系统的用户进行管理,并可以直接浏览用户表中相关数据信息,如图 17-11 所示。

图 17-11 "用户管理"界面

注册新用户时,要验证该用户名是否存在,如果已存在,则提示用户该用户已注册;修改用户信息成功后,提示用户修改成功;删除信息时,要提示用户是否确认要删除,用户确认后再完成信息删除。

4. 分析设计相关类

用户管理界面操作中涉及的主要类及类间关系,用 UML 图描述如图 17-12 所示。

1) 数据库操作类

该类在前面 DBUtil 类的基础上添加通用的数据库操作的方法,增、删、改、更新记录的操作,如图 17-13 所示。

2) 用户类

该类封装了用户基本信息,通过 get×××()方法和 set×××()方法获取和设置用户

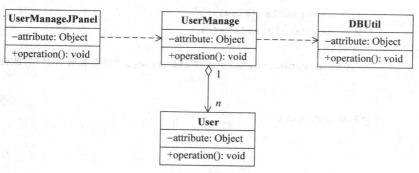

图 17-12　用户管理相关类 UML 图

基本信息，如图 17-14 所示。

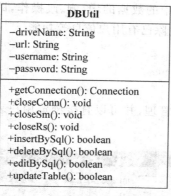

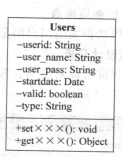

图 17-13　"数据库操作类"类图　　　　图 17-14　"用户类"类图

3）用户管理界面类

用户管理界面设计，如图 17-15 所示。

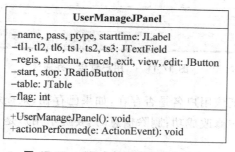

图 17-15　"用户管理界面类"类图

4）用户管理类

用户管理操作使用类，完成对图书管理系统操作用户的信息添加、修改、删除、显示等操作，类图如图 17-16 所示。

5．相关类主要代码

该模块设计涉及类的主要代码如下。

Users 类的代码如下：

UserManage
–jop: JOptionPane
+addUsers(User_name: String, User_pass: String, Type: String, date: String, flag: int): boolean +updateuser(User_name: String, User_pass: String, Type: String, date: String, flag: int): boolean +deleteUser(name: String): boolean +isExist(name: String): boolean +view(): void

图 17-16　"用户管理类"类图

```java
package com.ytvc.zjl.entity;
import java.util.Date;
public class Users {
    private String userid;
    private String user_name;
    private String user_pass;
    private Date startdate;
    private boolean valid;
    private String type;
    public String getUserid() {
        return userid;
    }
    public void setUserid(String userid) {
        this.userid=userid;
    }
    public String getUser_name() {
        return user_name;
    }
    public void setUser_name(String user_name) {
        this.user_name=user_name;
    }
    public String getUser_pass() {
        return user_pass;
    }
    public void setUser_pass(String user_pass) {
        this.user_pass=user_pass;
    }
    public Date getStartdate() {
        return startdate;
    }
    public void setStartdate(Date startdate) {
        this.startdate=startdate;
    }
    public boolean isValid() {
        return valid;
    }
```

```java
    public void setValid(boolean valid) {
        this.valid=valid;
    }
    public String getType() {
        return type;
    }
    public void setType(String type) {
        this.type=type;
    }
}
```

用户管理界面类 UserManageJPanel 的代码如下：

```java
package com.ytvc.zjl.user;
import javax.swing.*;
import com.ytvc.zjl.db.DBUtil;
import java.awt.*;
import java.awt.event.*;
import java.sql.*;
import java.util.Enumeration;
public class UserManageJPanel extends JFrame implements ActionListener {
    JLabel name,pass,ptype,starttime;
    JButton regis,shanchu,cancel,exit,view,edit;
    JTextField tl1,tl2,tl6,ts1,ts2,ts3;
    JPanel p,pp1,pp2,cc1;
    JScrollPane js;
    JComboBox type;
    JRadioButton start,stop;
    ButtonGroup bg;
    public static String User_name,User_pass,Type,date,valid;
    int flag;
    JTable table;
    public static int i=0,ii=0;
    public UserManageJPanel1() {
        name=new JLabel("用户名");
        pass=new JLabel("密 码");
        ptype=new JLabel("用户类型");
        type=new JComboBox();
        type.addItem("系统管理员");
        type.addItem("图书管理员");
        starttime=new JLabel("启用日期");
        start=new JRadioButton("启用");
        stop=new JRadioButton("停用");
        bg=new ButtonGroup();
        bg.add(start);
        bg.add(stop);
```

```
regis=new JButton("注册");
regis.addActionListener(this);
shanchu=new JButton("删除");
shanchu.addActionListener(this);
edit=new JButton("修改");
edit.addActionListener(this);
view=new JButton("浏览");
view.addActionListener(this);
cancel=new JButton("取消");
cancel.addActionListener(this);
exit=new JButton("退出");
exit.addActionListener(this);
tl1=new JTextField(10);
tl2=new JTextField(10);
tl6=new JTextField(10);
JPanel p1=new JPanel();
ts1=new JTextField(20);
ts1.setEditable(false);
ts2=new JTextField(10);
ts3=new JTextField(10);
p=new JPanel(new FlowLayout());
cc1=new JPanel(new BorderLayout());
pp2=new JPanel();
Container con=getContentPane();
con.setLayout(new FlowLayout());
pp2.add(name);
pp2.add(tl1);
pp2.add(pass);
pp2.add(tl2);
pp2.add(starttime);
pp2.add(tl6);
pp2.add(ptype);
pp2.add(type);
pp2.add(start);
pp2.add(stop);
p1.add(pp2);
p.add(regis);
p.add(edit);
p.add(view);
p.add(shanchu);
p.add(cancel);
p.add(exit);
String ta[]={ "用户名","用户类型","启用日期","状态" };
Object a[][]=new Object[30][4];
for (int i=0; i<=1; i++) {
```

```java
            for (int j=0; j<=1; j++) {
                a[i][j]="";
            }
        }
        table=new JTable(a,ta);
        js2=new JScrollPane(table);
        JPanel pjs=new JPanel();
        pjs.add(js2);
        con.add(p1);
        con.add(p);
        con.add(pjs);
        setBounds(240,250,780,318);
        setVisible(true);
        setTitle("用户管理");
    }
    public void actionPerformed(ActionEvent e) {
        User_name=tl1.getText();
        User_pass=tl2.getText();
        date=tl6.getText();
        Type=type.getSelectedItem().toString();
        flag=0;
        Enumeration<AbstractButton>radioBtns=bg.getElements();
        while (radioBtns.hasMoreElements()) {
            AbstractButton btn=radioBtns.nextElement();
            if (btn.isSelected()) {
                valid=btn.getText();
                if (valid.equals("启用")) {
                 flag=1;
                }
                break;
            }
        }
        if (e.getSource()==regis) {
        if (new UserManage().addUsers(User_name,User_pass,Type,date,flag)) {
            int j=0;
            table.setValueAt(User_name,i,j++);
            table.setValueAt(Type,i,j++);
            table.setValueAt(date,i,j++);
            table.setValueAt(valid,i,j++);
            }
        }
        if (e.getSource()==edit) {
            if (User_name.equals("")) {
                User_name=(String) table.getValueAt(table.getSelectedRow(),0);
            }
```

```
            if(new UserManage().updateUser(User_name,User_pass,Type,date,flag)){
                int j=0;
                table.setValueAt(User_name,i,j++);
                table.setValueAt(Type,i,j++);
                table.setValueAt(date,i,j++);
                table.setValueAt(valid,i,j++);
            }
        }
        if (e.getSource()==shanchu) {
            int k=0;
            int r=table.getSelectedRow();
            String ss=(String) table.getValueAt(r,0);
            if (new UserManage().deleteUser(ss)) {
                ResultSet result=null;
                String Valid;
                result=new UserManage().view();
                try {
                    while (result.next()) {
                        table.setValueAt(result.getString("user_name"),k,0);
                        table.setValueAt(result.getString("user_type"),k,1);
                        table.setValueAt(result.getString("startdate"),k,2);
                        if (result.getBoolean("valid")==true)
                            Valid="启用";
                        else
                            Valid="停用";
                        table.setValueAt(Valid,ii,3);
                        k++;
                    }
                } catch (SQLException e1) {
                    e1.printStackTrace();
                }
            }
        }
        if (e.getSource()==view) {
            ResultSet result=null;
            String Valid;
            result=new UserManage().view();
            try {
                while (result.next()) {
                    table.setValueAt(result.getString("user_name"),ii,0);
                    table.setValueAt(result.getString("user_type"),ii,1);
                    table.setValueAt(result.getString("startdate"),ii,2);
                    if (result.getBoolean("valid")==true)
                        Valid="启用";
                    else
```

```
                    Valid="停用";
                    table.setValueAt(Valid,ii,3);
                    ii++;
                }
            } catch (SQLException e1) {
                e1.printStackTrace();
            }
        }
        if (e.getSource()==cancel) {
            dispose();
            new com.ytvc.zjl.user.Login();
        }
        if (e.getSource()==exit) {
            System.exit(0);
        }
    }
}
```

用户管理类 UserManage 的主要方法代码如下:

```
//添加用户信息
    public boolean addUsers(String User_name,String User_pass,String Type,
            String date,int flag) {
        int row=0;
            Connection con=DBUtil.getConnection();
        if (!new UserManage1().isExist(User_name)) {
            try {
                Statement sm=con.createStatement();
        String sql="insert into userInfo(user_name,user_pass,user_type,
                startdate,valid) values('"+User_name+"','"+User_pass+"',
                '"+Type+"','"+date+"','"+flag+")";
                row=sm.executeUpdate(sql);
            } catch (SQLException e) {
                e.printStackTrace();
            }
        } else {
            JOptionPane.showMessageDialog(jop,"该用户已存在!!!");
            return false;
        }
        if (row>0) {
            JOptionPane.showMessageDialog(jop,"添加成功!");
            return true;
        } else {
            JOptionPane.showMessageDialog(jop,"添加失败!");
            return false;
        }
```

```java
    }
//修改用户信息
public boolean updateUser(String User_name,String User_pass,String Type,
        String date,int flag){
    int row=0;
    Connection con=DBUtil.getConnection();
    if (new UserManage1().isExist(User_name)) {
    try{
        Statement sm=con.createStatement();
        String sql="update userInfo set user_name='"+User_name+"',
                user_pass='"+User_pass+"',startdate='"+date+"',
                valid='"+flag+"' where user_name='"+User_name+"'";
        row=sm.executeUpdate(sql);
    } catch (SQLException e) {
        e.printStackTrace();
    }
    }
    else
    {JOptionPane.showMessageDialog(jop,"用户名不存在,请注册一个!!");
     return false;
    }
    if(row>0){
        JOptionPane.showMessageDialog(jop,"修改成功!");
        return true;
    }else{
        JOptionPane.showMessageDialog(jop,"修改失败!");
        return false;
    }
}
//删除用户信息
public boolean deleteUser(String name){
    int row=0;
    Connection con=DBUtil.getConnection();
    try {
        String sql="delete from userInfo where user_name='"+name+"'";
        Statement sm=con.createStatement();
        row=sm.executeUpdate(sql);
    } catch (SQLException e) {
        e.printStackTrace();
    }finally{
        try {
            con.close();
        } catch (SQLException e) {
            e.printStackTrace();
        }
```

```java
        }
        if(row>0){
            JOptionPane.showMessageDialog(jop,"删除成功!");
            return true;
        }else{
            JOptionPane.showMessageDialog(jop,"删除失败!");
            return false;
        }
    }
    //判断是否存在某个用户
    public boolean isExist(String name){
        Connection con=DBUtil.getConnection();
        try {
            String sql="select user_name from userinfo where user_name='"+name+"'";
            Statement sm=con.createStatement();
            ResultSet rs=sm.executeQuery(sql);
            if(rs.next()){
                return true;
            }
        } catch (SQLException e) {
            e.printStackTrace();
        }finally{
            try {
                con.close();
            } catch (SQLException e) {
                e.printStackTrace();
            }
        }
        return false;
    }
    //浏览数据库中的用户信息
    public ResultSet view(){
        Connection con=DBUtil.getConnection();
        ResultSet rs=null;
        Statement sm;
        String sql="select user_name,user_type,startdate,valid from userInfo";
        try {
            sm=con.createStatement();
            rs=sm.executeQuery(sql);
        }catch (SQLException e2) {

            e2.printStackTrace();
        }
        return rs;
    }
}
```

17.2.3 用户密码管理模块设计

1. 功能描述与分析

如果系统管理员/图书管理员发现密码不安全可以修改该管理员的密码,前提条件是管理员已进入到该系统中,要求用户输入以前密码,输入两次新密码,只有当原来密码正确,两次新密码相同,才能完成用户密码修改。

2. 涉及的数据库表

该功能模块使用用户信息表 userInfo。

3. 界面设计分析

通过如图 17-17 所示界面,修改用户信息表中用户密码信息。

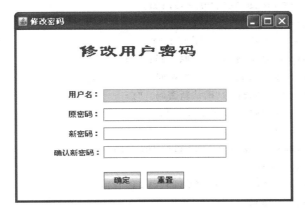

图 17-17 "修改用户密码"界面

4. 分析设计相关类

利用前面的 CheckUser 类、DBUtil 类,实现用户密码修改,UML 类图如图 17-18 所示。

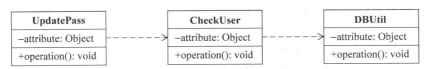

图 17-18 用户密码管理相关类 UML 图

"用户修改密码类"类图如图 17-19 所示。

5. 相关类主要代码

该模块设计涉及类的主要代码如下:

```
package com.ytvc.zjl.user;
import java.awt.Color;
import java.awt.Font;
import java.awt.event.*;
import javax.swing.*;
import com.ytvc.zjl.MI.MainInterface;
public class UpdatePass extends JPanel
```

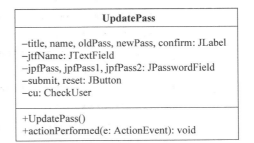

图 17-19 "用户修改密码类"类图

```java
implements ActionListener {
    JLabel title,name,oldPass,newPass,confirm;
    JTextField jtfName;
    JPasswordField jpfPass,jpfPass1,jpfPass2;
    JButton submit,reset;
    Box box,box1,box2;
    CheckUser cu=new CheckUser();
    public UpdatePass() {
        title=new JLabel("修改用户密码");
        title.setForeground(Color.blue);
        title.setFont(new Font("隶书",Font.BOLD,30));
        title.setBounds(100,20,300,30);
        add(title);
        name=new JLabel("    用户名：");
        oldPass=new JLabel("    原密码：");
        newPass=new JLabel("    新密码：");
        confirm=new JLabel("确认新密码：");
        box=Box.createVerticalBox();
        box.add(name);
        box.add(Box.createVerticalStrut(15));
        box.add(oldPass);
        box.add(Box.createVerticalStrut(15));
        box.add(newPass);
        box.add(Box.createVerticalStrut(15));
        box.add(confirm);
        box.setBounds(60,100,100,120);
        add(box);
        //自动获得用户登录时的用户名
        jtfName=new JTextField(MainInterface.status);
        jtfName.setEditable(false);
        jpfPass=new JPasswordField();
        jpfPass1=new JPasswordField();
        jpfPass2=new JPasswordField();
        box1=Box.createVerticalBox();
        box1.add(jtfName);
        box1.add(Box.createVerticalStrut(10));
        box1.add(jpfPass);
        box1.add(Box.createVerticalStrut(10));
        box1.add(jpfPass1);
        box1.add(Box.createVerticalStrut(10));
        box1.add(jpfPass2);
        box1.setBounds(140,100,200,120);
        add(box1);
        submit=new JButton("确定");
```

```
        reset=new JButton("重置");
        submit.addActionListener(this);
        reset.addActionListener(this);
        box2=Box.createHorizontalBox();
        box2.add(submit);
        box2.add(Box.createHorizontalStrut(10));
        box2.add(reset);
        box2.setBounds(140,240,150,30);
        add(box2);
        setLayout(null);
        setBackground(Color.white);
    }
    public void actionPerformed(ActionEvent e) {
        if (e.getSource()==submit) {
            String pass1=String.valueOf(jpfPass1.getPassword());
            String pass2=String.valueOf(jpfPass2.getPassword());
            if (cu.check(jtfName.getText(),jpfPass.getPassword())) {
                if (pass1.equals(pass2)) {
                    cu.updatePass(jtfName.getText(),pass1);
                } else {
                    JOptionPane.showMessageDialog(new JOptionPane(),
                        "两次密码输入不一致!");
                }
            } else {
                JOptionPane.showMessageDialog(new JOptionPane(),"密码有误!");
            }
            init();
        }
        if (e.getSource()==reset) {
            init();
        }
    }
    public void init() {
        jpfPass.setText("");
        jpfPass1.setText("");
        jpfPass2.setText("");
    }
}
```

17.2.4 读者信息管理模块设计

1. 功能描述与分析

图书管理员通过该模块对图书馆中的读者进行信息登记,完成对读者基本信息的录入、修改和删除操作。

2. 涉及的数据库表

读者信息管理模块涉及读者信息表 reader、读者类型表 readertype、部门信息表 dept，表结构如表 17-2～表 17-4 所示。

表 17-2　读者信息表结构

字段名	数据类型	长度	允许空	是否主键	含　义
reader_id	int	4B	不允许	是	读者编号
reader_name	varchar	20 字符	不允许		读者名
reader_sex	varchar	1 字符	允许		读者性别
reader_special	varchar	20 字符	允许		读者专业
reader_class	varchar	20 字符	允许		读者班级
reader_tel	varchar	20 字符	允许		读者电话
reader_begin	datetime	8B	允许		启用日期
reader_state	bit	1B	允许		使用状态
dept_id	int	4B	不允许		读者部门编号
reader_typeid	int	4B	不允许		读者类型编号

表 17-3　读者类型表结构

字段名	数据类型	长度	允许空	是否主键	含　义
reader_typeid	int	4B	不允许	是	读者类型编号
reader_type	varchar	20 字符	不允许		读者类型
borrow_num	int	4B	允许		借阅数量
borrow_limit	int	4B	允许		借书期限（月）
fine_money	float	4B	允许		超期日罚金额
validtime	int	4B	允许		有效使用期

表 17-4　部门信息表结构

字段名	数据类型	长度	允许空	是否主键	含　义
dept_id	int	4B	不允许	是	部门编号
dept_name	varchar	20 字符	不允许		部门名称
dept_leader	varchar	20 字符	允许		部门负责人
dept_contact	varchar	20 字符	允许		部门联系人
dept_tel	varchar	20 字符	允许		联系电话

3. 界面设计分析

读者信息涉及读者的类型、所属部门等重要信息，因此要设置这些相关信息，进入到"读者类型设置"界面，如图 17-20 所示，设置读者类型；进入"部门信息设置"界面，如图 17-21 所示，设置部门信息。

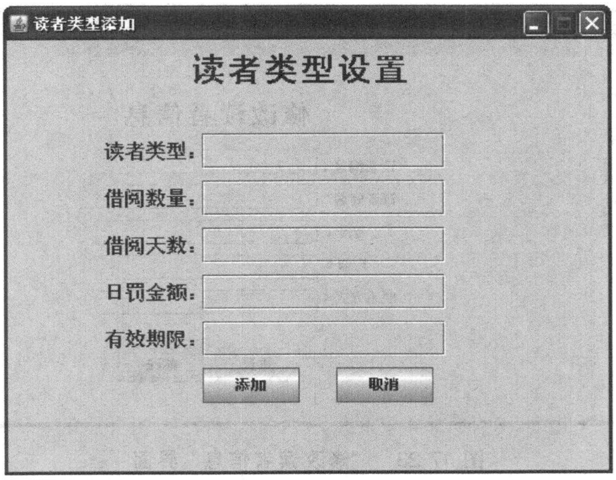

图 17-20 "读者类型设置"界面

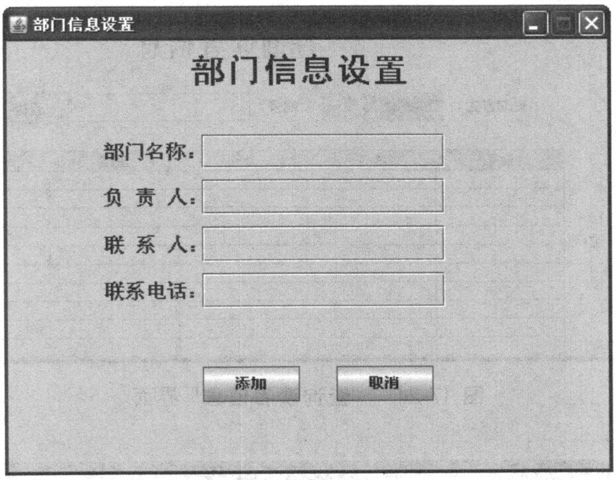

图 17-21 "部门信息设置"界面

完善基础数据的维护后,完成读者信息的维护,其界面如图 17-22～图 17-25 所示。

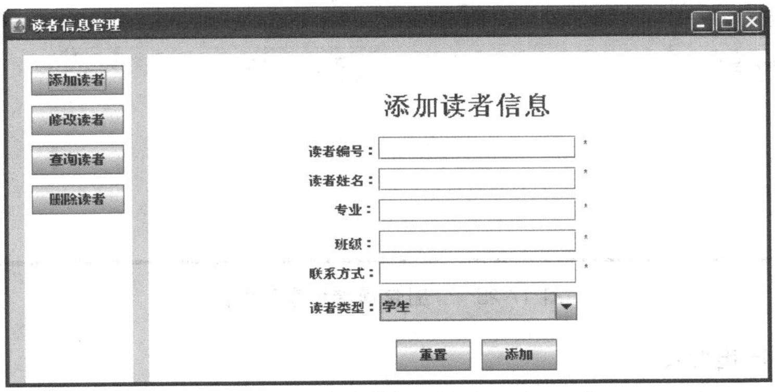

图 17-22 "添加读者信息"界面

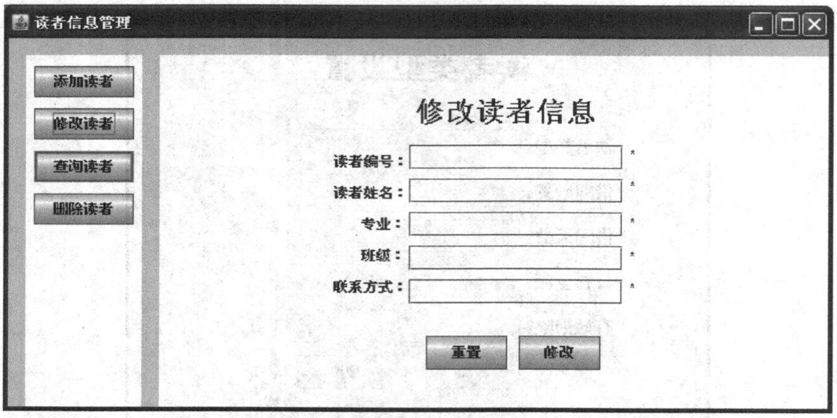

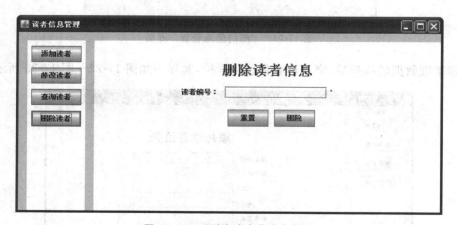

图 17-23 "修改读者信息"界面

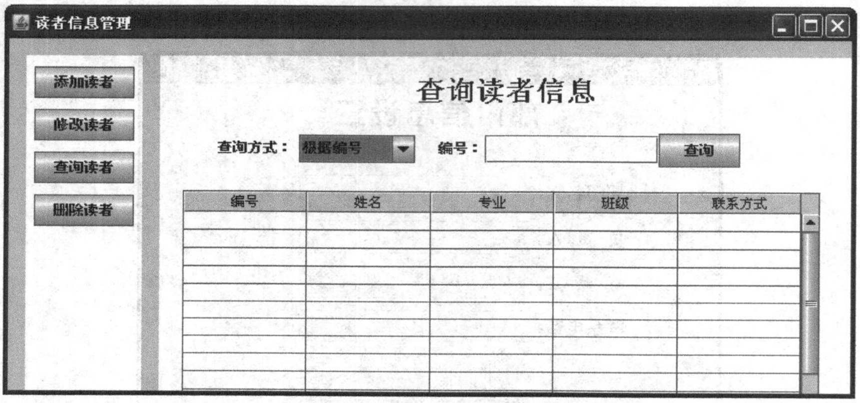

图 17-24 "查询读者信息"界面

图 17-25 "删除读者信息"界面

4. 分析设计相关类

读者管理界面操作中涉及的主要类及类间关系，用 UML 图描述如图 17-26 所示。

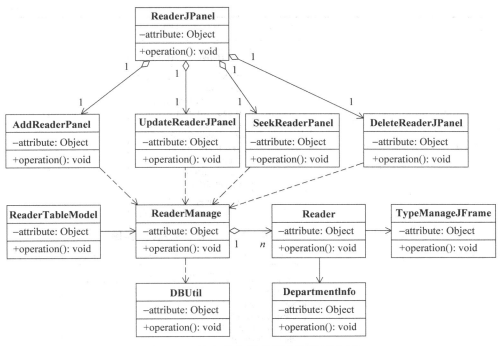

图 17-26 读者信息管理相关类间 UML 图

使用 DBUtil 类与数据库连接,对读者表中信息进行操作,完成对读者的管理,各个类描述如表 17-5 所示。各个类所对应的类图如图 17-27～图 17-36 所示。

表 17-5 读者信息管理相关类列表

类　名	UML 类图
读者类型设置类	**TypeManageJFrame** –lTypeName: JLabel –jtfTypeName: JTextField –jtfNum: JTextField –ltfNum: JLabel –jtfDay: JTextField –lDay: JLabel –jtfMoney: JTextField –lMoney: JLabel –jtfLast: JTextField –lLast: JLabel –submint: JButton –cancel: JButton +TypeManageJFrame() +actionPerformed(): void 图 17-27 "读者类型类" 类图

续表

类 名	UML 类图
部门信息类	**DepartmentInfo** -jtfDeptName: JTextField -lDeptName: JLabel -jtfLeader: JTextField -lLeader: JLabel -jtfLinkman: JTextField -lLinkman: JLabel -jtfTel: JTextField -lTel: JLabel -sumbit: JButton -cancel: JButton +DepartmentInfo() +actionPerformed(): void 图 17-28　"部门信息类"类图
添加读者信息类	**AddReader JPanel** –jtfID: JTextField –jtfName: JTextField –jtfClass: JTextField –jftSpecial: JTextField –jtfPhon: JTextField –jIID, jIName, jIClass, jISpecial, jIPhone, jIType: JLabel –submit, cancle: JButton +AddReaderJPanel() +actionPerformed(): void +init(): void +focusLost(): void 图 17-29　"添加读者信息类"类图
修改读者信息类	**UpdateReaderJPanel** –jtfID, jtfName, jtfClass, jtfSpecial, jtfPhon: JTextField –jIID, jIName, jIClass, jISpecial, jIPhone: JLabel –submint, cancel: JButton +UpdateReaderJPanel() +actionPerformed(): void +init(): void focusLost(): void 图 17-30　"修改读者信息类"类图
查询读者信息类	**SeekReaderJPanel** –jtfID: JTextField –lID: JLabel –jbSeek: JButton +SeekReaderJPanel() +actionPerformed(): void +init(): void +focusLost(): void 图 17-31　"查询读者信息类"类图

续表

类 名	UML 类图
删除读者信息类	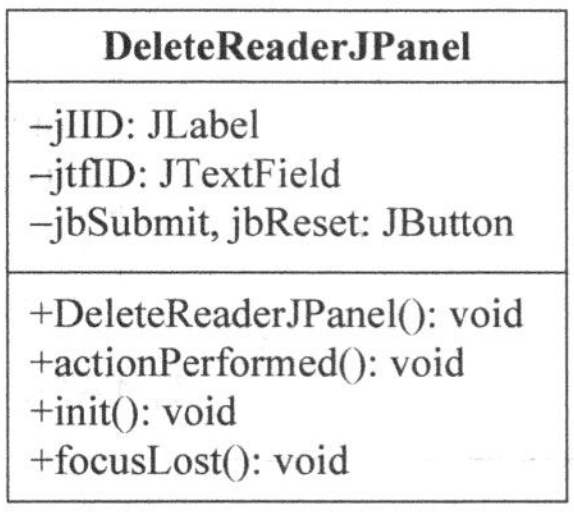 **DeleteReaderJPanel** –jlID: JLabel –jtfID: JTextField –jbSubmit, jbReset: JButton +DeleteReaderJPanel(): void +actionPerformed(): void +init(): void +focusLost(): void 图 17-32　"删除读者信息类"类图
读者信息管理类	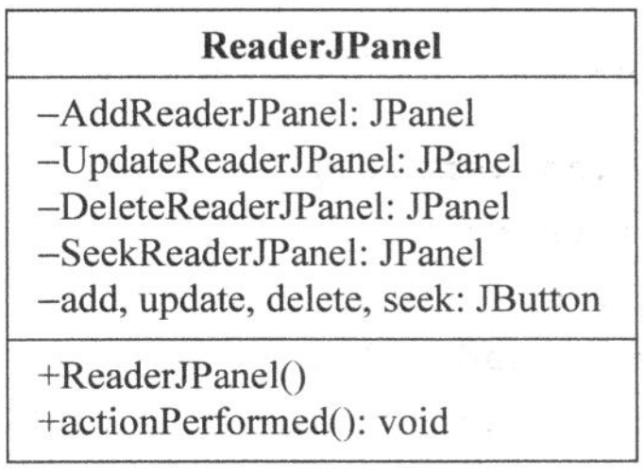 **ReaderJPanel** –AddReaderJPanel: JPanel –UpdateReaderJPanel: JPanel –DeleteReaderJPanel: JPanel –SeekReaderJPanel: JPanel –add, update, delete, seek: JButton +ReaderJPanel() +actionPerformed(): void 图 17-33　"读者信息管理类"类图
读者实体类	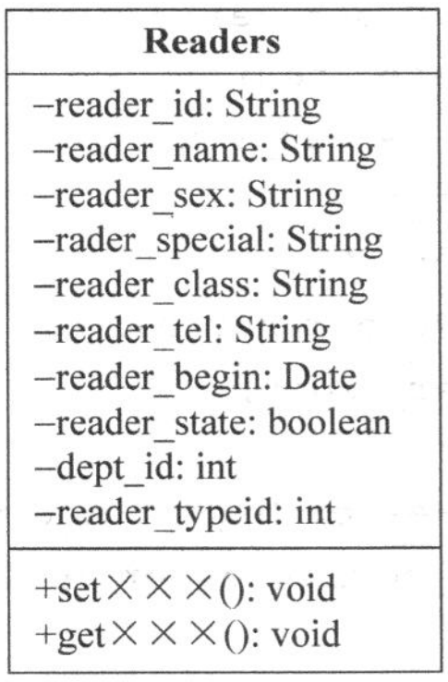 **Readers** –reader_id: String –reader_name: String –reader_sex: String –rader_special: String –reader_class: String –reader_tel: String –reader_begin: Date –reader_state: boolean –dept_id: int –reader_typeid: int +set×××(): void +get×××(): void 图 17-34　"读者实体类"类图
读者信息操作类	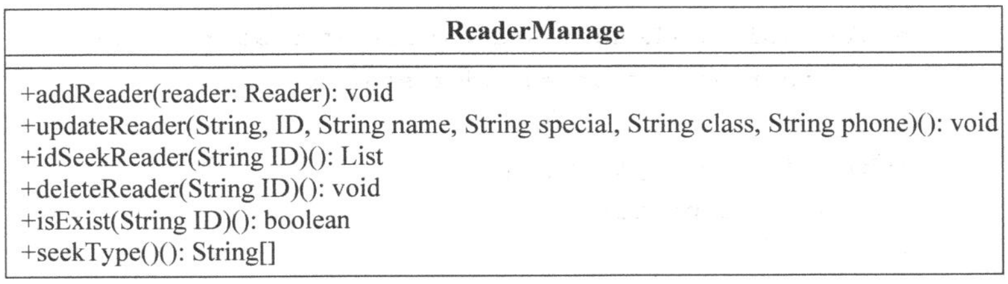 **ReaderManage** +addReader(reader: Reader): void +updateReader(String, ID, String name, String special, String class, String phone)(): void +idSeekReader(String ID)(): List +deleteReader(String ID)(): void +isExist(String ID)(): boolean +seekType()(): String[] 图 17-35　"读者信息操作类"类图

类　　名	UML 类图
读者信息表模型类	**Reader TableModel** −rm: ReaderManage −list: List +ReaderTableModel() +ReaderTableModel(String ID)()

图 17-36　"读者信息表模型类"类图

5．相关类主要代码

该模块设计涉及的类主要代码如下。

读者实体类 Readers：

```java
package com.ytvc.zjl.entity;
import java.util.Date;
public class Readers {
    private String reader_id;
    private String reader_name;
    private String reader_sex;
    private String reader_special;
    private String reader_class;
    private String reader_tel;
    private Date reader_begin;
    private boolean reader_state;
    private int dept_id;
    private int reader_typeid;
    public Date getReader_begin() {
        return reader_begin;
    }
    public void setReader_begin(Date reader_begin) {
        this.reader_begin=reader_begin;
    }
    public boolean isReader_state() {
        return reader_state;
    }
    public void setReader_state(boolean reader_state) {
        this.reader_state=reader_state;
    }
    public int getDept_id() {
        return dept_id;
    }
    public void setDept_id(int dept_id) {
        this.dept_id=dept_id;
    }
    public int getReader_typeid() {
```

```java
        return reader_typeid;
    }
    public void setReader_typeid(int reader_typeid) {
        this.reader_typeid=reader_typeid;
    }
    public String getReader_class() {
        return reader_class;
    }
    public void setReader_class(String reader_class) {
        this.reader_class=reader_class;
    }
    public String getReader_id() {
        return reader_id;
    }
    public void setReader_id(String reader_id) {
        this.reader_id=reader_id;
    }
    public String getReader_name() {
        return reader_name;
    }
    public void setReader_name(String reader_name) {
        this.reader_name=reader_name;
    }
    public String getReader_sex() {
        return reader_sex;
    }
    public void setReader_sex(String reader_sex) {
        this.reader_sex=reader_sex;
    }
    public String getReader_special() {
        return reader_special;
    }
    public void setReader_special(String reader_special) {
        this.reader_special=reader_special;
    }
    public String getReader_tel() {
        return reader_tel;
    }
    public void setReader_tel(String reader_tel) {
        this.reader_tel=reader_tel;
    }
}
```

读者信息管理类 ReaderJPanel：

```java
public class ReaderJPanel extends JPanel implements ActionListener{
    JPanel jpLeft;
    JButton addReader,updateReader,deleteReader,seekReader;
```

```java
        public static AddReaderJPanel arj=new AddReaderJPanel();
        public static UpdateReaderJPanel urj=new UpdateReaderJPanel();
        public static DeleteReaderJPanel drj=new DeleteReaderJPanel();
        public static SeekReaderJPanel srj=new SeekReaderJPanel();
        public ReaderJPanel() {
            arj.setVisible(false);
            urj.setVisible(false);
            drj.setVisible(false);
            srj.setVisible(false);
            setLayout(null);
            jpLeft=new JPanel();
            addReader=new JButton("添加读者");
            addReader.addActionListener(this);
            updateReader=new JButton("修改读者");
            updateReader.addActionListener(this);
            seekReader=new JButton("查询读者");
            seekReader.addActionListener(this);
            deleteReader=new JButton("删除读者");
            deleteReader.addActionListener(this);
            jpLeft.add(Box.createHorizontalStrut(10));
            jpLeft.add(addReader);
            jpLeft.add(Box.createHorizontalStrut(10));
            jpLeft.add(Box.createHorizontalStrut(10));
            jpLeft.add(Box.createHorizontalStrut(10));
            jpLeft.add(updateReader);
            jpLeft.add(Box.createHorizontalStrut(10));
            jpLeft.add(Box.createHorizontalStrut(10));
            jpLeft.add(Box.createHorizontalStrut(10));
            jpLeft.add(seekReader);
            jpLeft.add(Box.createHorizontalStrut(10));
            jpLeft.add(Box.createHorizontalStrut(10));
            jpLeft.add(Box.createHorizontalStrut(10));
            jpLeft.add(deleteReader);
            jpLeft.setBounds(15,15,100,370);
            jpLeft.setBackground(Color.white);
            add(jpLeft);
            arj.setBounds(130,15,685,370);
            add(arj);
            urj.setBounds(130,15,685,370);
            add(urj);
            drj.setBounds(130,15,685,370);
            add(drj);
            srj.setBounds(130,15,685,370);
            add(srj);
        }
```

```java
    public void actionPerformed(ActionEvent e) {
        if(e.getSource()==addReader){
            arj.comboxInit();
            MainInterface.jp.setVisible(false);
            urj.setVisible(false);
            drj.setVisible(false);
            srj.setVisible(false);
            arj.setVisible(true);
            arj.init();
        }
        if(e.getSource()==updateReader){
            MainInterface.jp.setVisible(false);
            arj.setVisible(false);
            drj.setVisible(false);
            srj.setVisible(false);
            urj.setVisible(true);
            urj.init();
        }
        if(e.getSource()==seekReader){
            MainInterface.jp.setVisible(false);
            arj.setVisible(false);
            urj.setVisible(false);
            drj.setVisible(false);
            srj.setVisible(true);
            srj.init();
        }
        if(e.getSource()==deleteReader){
            MainInterface.jp.setVisible(false);
            arj.setVisible(false);
            urj.setVisible(false);
            srj.setVisible(false);
            drj.setVisible(true);
            drj.init();
        }
    }
}
```

添加读者信息类 AddReaderJPanel：

```java
package com.ytvc.zjl.reader;
import java.awt.*;
import java.awt.event.*;
import javax.swing.*;
import com.ytvc.zjl.entity.Readers;
import com.ytvc.zjl.reader.impl.ReaderManage;
public class AddReaderJPanel extends JPanel implements ActionListener,
```

```
FocusListener,ItemListener{
    Box box1,box2,box3,boxBase;
    JLabel jl,jlID,jlName,jlClass,jlSpecial,jlPhone,jlType,jl1,jl2,jl3,jl4,jl5;
    JButton jbSubmit,jbReset;
    JTextField jtfID,jtfName,jtfClass,jtfSpecial,jtfPhon;
    JComboBox comBox;
    ReaderManage rm=new ReaderManage();
    public AddReaderJPanel() {
        jl1=new JLabel(" * ");
        jl1.setForeground(Color.red);
        jl2=new JLabel(" * ");
        jl2.setForeground(Color.red);
        jl3=new JLabel(" * ");
        jl3.setForeground(Color.red);
        jl4=new JLabel(" * ");
        jl4.setForeground(Color.red);
        jl5=new JLabel(" * ");
        jl5.setForeground(Color.red);
        jl=new JLabel("添加读者信息");
        jl.setFont(new Font("宋体",Font.BOLD,25));
        jl.setForeground(Color.blue);
        jl.setBounds(220,10,200,80);
        add(jl);
        jlID=new JLabel("读者编号：");
        jlName=new JLabel("读者姓名：");
        jlSpecial=new JLabel("    专业：");
        jlClass=new JLabel("    班级：");
        jlPhone=new JLabel("联系方式：");
        jlType=new JLabel("读者类型：");
        jbSubmit=new JButton("添加");
        jbReset=new JButton("重置");
        jbSubmit.addActionListener(this);
        jbReset.addActionListener(this);
        jtfID=new JTextField();
        jtfName=new JTextField();
        jtfSpecial=new JTextField();
        jtfClass=new JTextField();
        jtfPhon=new JTextField();
        comBox=new JComboBox(rm.seekType());
        jtfID.addFocusListener(this);
        jtfName.addFocusListener(this);
        jtfSpecial.addFocusListener(this);
        jtfClass.addFocusListener(this);
        jtfPhon.addFocusListener(this);
        comBox.addItemListener(this);
```

```java
box1=Box.createVerticalBox();
box1.add(jlID);
box1.add(Box.createVerticalStrut(10));
box1.add(jlName);
box1.add(Box.createVerticalStrut(10));
box1.add(jlSpecial);
box1.add(Box.createVerticalStrut(15));
box1.add(jlClass);
box1.add(Box.createVerticalStrut(10));
box1.add(jlPhone);
box1.add(Box.createVerticalStrut(15));
box1.add(jlType);
box2=Box.createVerticalBox();
box2.add(jtfID);
box2.add(Box.createVerticalStrut(8));
box2.add(jtfName);
box2.add(Box.createVerticalStrut(8));
box2.add(jtfSpecial);
box2.add(Box.createVerticalStrut(8));
box2.add(jtfClass);
box2.add(Box.createVerticalStrut(8));
box2.add(jtfPhon);
box2.add(Box.createVerticalStrut(8));
box2.add(comBox);
box3=Box.createVerticalBox();
box3.add(jl1);
box3.add(Box.createVerticalStrut(10));
box3.add(jl2);
box3.add(Box.createVerticalStrut(15));
box3.add(jl3);
box3.add(Box.createVerticalStrut(12));
box3.add(jl4);
box3.add(Box.createVerticalStrut(10));
box3.add(jl5);
boxBase=Box.createHorizontalBox();
boxBase.add(box1);
boxBase.add(box2);
boxBase.setBounds(150,80,250,180);
add(boxBase);
box3.setBounds(405,80,250,142);
add(box3);
jbReset.setBounds(230,275,70,30);
add(jbReset);
jbSubmit.setBounds(310,275,70,30);
add(jbSubmit);
```

```java
        setBackground(Color.white);
        setLayout(null);
    }
    public void actionPerformed(ActionEvent e) {
        if(e.getSource()==jbReset){
            init();
        }
        if(e.getSource()==jbSubmit){
if (jtfID.getText().equals("")||jtfName.getText().equals("")||jtfSpecial.
  getText().equals("")||jtfClass.getText().equals("")||jtfPhon.getText()
  .equals("")){
            }else{
                Readers reader=new Readers();
                reader.setReader_id(jtfID.getText());
                reader.setReader_name(jtfName.getText());
                reader.setReader_special(jtfSpecial.getText());
                reader.setReader_class(jtfClass.getText());
                reader.setReader_tel(jtfPhon.getText());
                reader.setReader_typeid(rm.seekTypeID((String)comBox
                            .getSelectedItem()));rm.addReader(reader);
                init();
            }
        }
    }
    public void focusGained(FocusEvent e) {}
    public void focusLost(FocusEvent e) {
        if(e.getSource()==jtfID){
            if(jtfID.getText().equals("")){
                jl1.setText("* 学号不可以为空");
            }else{
                if(rm.isExist(jtfID.getText())){
                    jl1.setText("* 此读者已存在");
                }else{
                    jl1.setText("* ");
                }
            }
        }
        if(e.getSource()==jtfName){
            if(jtfName.getText().equals("")){
                jl2.setText("* 姓名不可以为空");
            }else{
                jl2.setText("* ");
            }
```

```java
        if(e.getSource()==jtfSpecial){
            if(jtfSpecial.getText().equals("")){
                jl3.setText("* 专业不可以为空");
            }else{
                jl3.setText("*");
            }
        }
        if(e.getSource()==jtfClass){
            if(jtfClass.getText().equals("")){
                jl4.setText("* 班级不可以为空");
            }else{
                jl4.setText("*");
            }
        }
        if(e.getSource()==jtfPhon){
            if(jtfPhon.getText().equals("")){
                jl5.setText("* 联系方式不可以为空");
            }else{
                jl5.setText("*");
            }
        }
    }
    public void init(){
        jtfID.setText("");
        jtfName.setText("");
        jtfSpecial.setText("");
        jtfClass.setText("");
        jtfClass.setEditable(true);
        jtfPhon.setText("");
        comBox.setSelectedIndex(0);
        jl1.setText("*");
        jl2.setText("*");
        jl3.setText("*");
        jl4.setText("*");
        jl5.setText("*");
    }
    public void comboxInit(){
        if(comBox.getItemCount()!=0){
            comBox.removeAllItems();
        }
        String[] items=rm.seekType();
        for(int i=0;i<items.length;i++){
            comBox.addItem(items[i]);
        }
    }
```

```java
    public void itemStateChanged(ItemEvent e){
        if(e.getSource()==comBox){
            if(comBox.getItemCount()!=0){
                if(!comBox.getSelectedItem().equals("学生")){
                    jtfClass.setText("null");
                    jtfClass.setEditable(false);
                }
            }
        }
    }
}
```

修改读者信息类 UpdateReaderJPanel：

```java
package com.ytvc.zjl.reader;
import java.awt.*;
import java.awt.event.*;
import javax.swing.*;
import com.ytvc.zjl.entity.Readers;
import com.ytvc.zjl.reader.impl.ReaderManage;
public class UpdateReaderJPanel extends JPanel implements ActionListener,FocusListener{
    Box box1,box2,box3,boxBase;
    JLabel jl,jlID,jlName,jlClass,jlSpecial,jlPhone,jl1,jl2,jl3,jl4,jl5;
    JButton jbSubmit,jbReset;
    JTextField jtfID,jtfName,jtfClass,jtfSpecial,jtfPhon;
    Readers reader;
    ReaderManage rm=new ReaderManage();
    public UpdateReaderJPanel() {
        jl1=new JLabel(" * ");
        jl1.setForeground(Color.red);
        jl2=new JLabel(" * ");
        jl2.setForeground(Color.red);
        jl3=new JLabel(" * ");
        jl3.setForeground(Color.red);
        jl4=new JLabel(" * ");
        jl4.setForeground(Color.red);
        jl5=new JLabel(" * ");
        jl5.setForeground(Color.red);
        jl=new JLabel("修改读者信息");
        jl.setFont(new Font("宋体",Font.BOLD,25));
        jl.setForeground(Color.blue);
        jl.setBounds(220,10,200,80);
        add(jl);
        jlID=new JLabel("读者编号：");
        jlName=new JLabel("读者姓名：");
```

```
jlSpecial=new JLabel("    专业：");
jlClass=new JLabel("    班级：");
jlPhone=new JLabel("联系方式：");
jbSubmit=new JButton("修改");
jbReset=new JButton("重置");
jbSubmit.addActionListener(this);
jbReset.addActionListener(this);
jtfID=new JTextField();
jtfName=new JTextField();
jtfSpecial=new JTextField();
jtfClass=new JTextField();
jtfPhon=new JTextField();
jtfID.addFocusListener(this);
jtfName.addFocusListener(this);
jtfSpecial.addFocusListener(this);
jtfClass.addFocusListener(this);
jtfPhon.addFocusListener(this);
box1=Box.createVerticalBox();
box1.add(jlID);
box1.add(Box.createVerticalStrut(10));
box1.add(jlName);
box1.add(Box.createVerticalStrut(10));
box1.add(jlSpecial);
box1.add(Box.createVerticalStrut(10));
box1.add(jlClass);
box1.add(Box.createVerticalStrut(10));
box1.add(jlPhone);
box2=Box.createVerticalBox();
box2.add(jtfID);
box2.add(Box.createVerticalStrut(8));
box2.add(jtfName);
box2.add(Box.createVerticalStrut(8));
box2.add(jtfSpecial);
box2.add(Box.createVerticalStrut(8));
box2.add(jtfClass);
box2.add(Box.createVerticalStrut(8));
box2.add(jtfPhon);
box3=Box.createVerticalBox();
box3.add(jl1);
box3.add(Box.createVerticalStrut(10));
box3.add(jl2);
box3.add(Box.createVerticalStrut(15));
box3.add(jl3);
box3.add(Box.createVerticalStrut(12));
box3.add(jl4);
```

```java
        box3.add(Box.createVerticalStrut(10));
        box3.add(jl5);
        boxBase=Box.createHorizontalBox();
        boxBase.add(box1);
        boxBase.add(box2);
        boxBase.setBounds(150,80,250,142);
        add(boxBase);
        box3.setBounds(405,80,250,142);
        add(box3);
        jbReset.setBounds(230,250,70,30);
        add(jbReset);
        jbSubmit.setBounds(310,250,70,30);
        add(jbSubmit);
        setBackground(Color.white);
        setLayout(null);
    }
    public void actionPerformed(ActionEvent e) {
        if(e.getSource()==jbReset){
            init();
        }
        if(e.getSource()==jbSubmit){
            if (jtfID.getText().equals("")||jtfName.getText().equals("")||
                jtfSpecial.getText().equals("")||jtfClass.getText()
                .equals("")||jtfPhon.getText().equals("")){
            }else{
                rm.updateReader(jtfID.getText(),jtfName.getText(),jtfSpecial
                .getText(),jtfClass.getText(),jtfPhon.getText());
                init();
            }
        }
    }
    public void focusGained(FocusEvent e) {}
    public void focusLost(FocusEvent e) {
        if(e.getSource()==jtfID){
            if(jtfID.getText().equals("")){
                jl1.setText("* 学号不可以为空");
            }else{
                if(rm.isExist(jtfID.getText())){
                    jl1.setText("*");
                    reader=(Readers)(rm.idSeekReader(jtfID.getText())).get(0);
                    jtfName.setText(reader.getReader_name());
                    jtfSpecial.setText(reader.getReader_special());
                    jtfClass.setText(reader.getReader_class());
                    jtfPhon.setText(reader.getReader_tel());
                }else{
```

```
                jl1.setText(" * 不存在此读者信息");
            }
        }
    }
    if(e.getSource()==jtfName){
        if(jtfName.getText().equals("")){
            jl2.setText(" * 姓名不可以为空");
        }else{
            jl2.setText(" * ");
        }
    }
    if(e.getSource()==jtfSpecial){
        if(jtfSpecial.getText().equals("")){
            jl3.setText(" * 专业不可以为空");
        }else{
            jl3.setText(" * ");
        }
    }
    if(e.getSource()==jtfClass){
        if(jtfClass.getText().equals("")){
            jl4.setText(" * 班级不可以为空");
        }else{
            jl4.setText(" * ");
        }
    }
    if(e.getSource()==jtfPhon){
        if(jtfPhon.getText().equals("")){
            jl5.setText(" * 联系方式不可以为空");
        }else{
            jl5.setText(" * ");
        }
    }
}
public void init(){
    jtfID.setText("");
    jtfName.setText("");
    jtfSpecial.setText("");
    jtfClass.setText("");
    jtfPhon.setText("");
    jl1.setText(" * ");
    jl2.setText(" * ");
    jl3.setText(" * ");
    jl4.setText(" * ");
    jl5.setText(" * ");
}
}
```

查询读者信息类 SeekReaderJPanel：

```java
public class SeekReaderJPanel extends JPanel implements ActionListener{
    JTable jt;
    JLabel jl,jl1,jlID;
    JTextField jtfID;
    JComboBox comBox;
    JButton jbSeek;
    JScrollPane js;
    public SeekReaderJPanel() {
        jl=new JLabel("查询读者信息");
        jl.setFont(new Font("宋体",Font.BOLD,25));
        jl.setForeground(Color.blue);
        jl.setBounds(220,0,200,60);
        add(jl);
        jl1=new JLabel("查询方式：");
        jl1.setBounds(50,70,70,25);
        add(jl1);
        comBox=new JComboBox();
        comBox.addItem("根据编号");
        comBox.addItem("查询全部");
        comBox.setBounds(120,70,100,25);
        comBox.addActionListener(this);
        add(comBox);
        jlID=new JLabel("编号：");
        jlID.setBounds(240,70,40,25);
        add(jlID);
        jtfID=new JTextField();
        jtfID.setBounds(280,70,150,25);
        add(jtfID);
        jbSeek=new JButton("查询");
        jbSeek.addActionListener(this);
        jbSeek.setBounds(430,68,70,30);
        add(jbSeek);
        jt=new JTable(new ReaderTableModel());
        js=new JScrollPane(jt);
        js.setBounds(20,120,550,240);
        add(js);
        setBackground(Color.white);
        setLayout(null);
    }
    public void actionPerformed(ActionEvent e) {
        if(e.getSource()==comBox){
            if(comBox.getSelectedIndex()==0){
                jtfID.setEditable(true);
            }
```

```java
            if(comBox.getSelectedIndex()==1){
                jtfID.setText("");
                jtfID.setEditable(false);
            }
        }
        if(e.getSource()==jbSeek){
            if(comBox.getSelectedIndex()==0&&jtfID.getText().equals("")){
            }else{
                jt.setModel(new ReaderTableModel(jtfID.getText()));
            }
        }
    }
    public void init(){
        jtfID.setText("");
        comBox.setSelectedIndex(0);
        jt.setModel(new ReaderTableModel());
    }
}
```

删除读者信息类 DeleteReaderJPanel:

```java
public class DeleteReaderJPanel extends JPanel implements ActionListener,FocusListener{
    Box box1,box2,box3,boxBase;
    JLabel jl,jlID,jl1;
    JButton jbSubmit,jbReset;
    JTextField jtfID;
    ReaderManage rm=new ReaderManage();
    public DeleteReaderJPanel() {
        jl1=new JLabel(" * ");
        jl1.setForeground(Color.red);
        jl=new JLabel("删除读者信息");
        jl.setFont(new Font("宋体",Font.BOLD,25));
        jl.setForeground(Color.blue);
        jl.setBounds(220,10,200,80);
        add(jl);
        jlID=new JLabel("读者编号:");
        jbSubmit=new JButton("删除");
        jbReset=new JButton("重置");
        jbSubmit.addActionListener(this);
        jbReset.addActionListener(this);
        jtfID=new JTextField();
        jtfID.addFocusListener(this);
        box1=Box.createHorizontalBox();
        box1.add(jlID);
```

```
        box1.add(Box.createHorizontalStrut(10));
        box1.add(jtfID);
        box2=Box.createVerticalBox();
        box2.add(jl1);
        box1.setBounds(150,80,250,20);
        add(box1);
        box2.setBounds(405,80,250,142);
        add(box2);
        jbReset.setBounds(230,120,70,30);
        add(jbReset);
        jbSubmit.setBounds(310,120,70,30);
        add(jbSubmit);
        setBackground(Color.white);
        setLayout(null);
    }
    public void actionPerformed(ActionEvent e){
        if(e.getSource()==jbReset){
            init();
        }
        if(e.getSource()==jbSubmit){
            if(jtfID.getText().equals("")){
                jl1.setText("* 学号不可以为空");
            }else{
                if(rm.isExist(jtfID.getText())){
                    jl1.setText("* ");
                    rm.deleteReader(jtfID.getText());
                    init();
                }else{
                    jl1.setText("* 不存在此读者信息");
                }
            }
        }
    }
    public void focusGained(FocusEvent e){}
    public void focusLost(FocusEvent e){
        if(e.getSource()==jtfID){
            if(jtfID.getText().equals("")){
                jl1.setText("* 学号不可以为空");
            }else{
                if(rm.isExist(jtfID.getText())){
                    jl1.setText("* ");
                }else{
                    jl1.setText("* 不存在此读者信息");
                }
            }
```

```
        }
    }
    public void init(){
        jtfID.setText("");
        jl1.setText(" * ");
    }
}
```

读者信息操作类 ReaderManage：

```
package com.ytvc.zjl.reader.impl;
import java.sql.*;
import java.util.*;
import javax.swing.JOptionPane;
import com.ytvc.zjl.db.DBUtil;
import com.ytvc.zjl.entity.Readers;
public class ReaderManage {
    JOptionPane jop=new JOptionPane();
    //添加读者
    public void addReader(Readers reader){
        int row=0;
        Connection con=DBUtil.getConnection();
        try {
            String sql="insert into reader(reader_id,reader_name,reader_
            typeid,reader_special,reader_class,reader_tel) values('"+reader
            .getReader_id()+"','"+reader.getReader_name()+"','"+reader
            .getReader_typeid()+"','"+reader.getReader_special()+"','"+reader
            .getReader_class()+"','"+reader.getReader_tel()+"')";
            Statement sm=con.createStatement();
            row=sm.executeUpdate(sql);
        } catch (SQLException e) {
            e.printStackTrace();
        }finally{
            try {
                con.close();
            } catch (SQLException e) {
                e.printStackTrace();
            }
        }
        if(row>0){
            JOptionPane.showMessageDialog(jop,"添加成功!");
        }else{
            JOptionPane.showMessageDialog(jop,"添加失败!");
        }
    }
    //修改读者
```

```java
public void updateReader(String ID,String name,String special,String class,
String phone){
    int row=0;
    Connection con=DBUtil.getConnection();
    try {
        String sql="update reader set reader_name='"+name+"',reader_
                special='"+special+"',reader_class='"+clas+"',
                reader_tel='"+phone+"' where reader_id='"+ID+"'";
        Statement sm=con.createStatement();
        row=sm.executeUpdate(sql);
    } catch (SQLException e) {
        e.printStackTrace();
    }finally{
        try {
            con.close();
        } catch (SQLException e) {
            e.printStackTrace();
        }
    }
    if(row>0){
        JOptionPane.showMessageDialog(jop,"修改成功!");
    }else{
        JOptionPane.showMessageDialog(jop,"修改失败!");
    }
}
//根据传过来的ID返回查询结果
public List idSeekReader(String ID){
    List list=new ArrayList();
    Readers reader;
    Connection con=DBUtil.getConnection();
    try {
        String sql;
        if(ID.equals("")){
            sql="select * from reader";
        }else{
            sql="select * from reader where reader_id='"+ID+"'";
        }
        Statement sm=con.createStatement();
        ResultSet rs=sm.executeQuery(sql);
        while(rs.next()){
            reader=new Readers();
            reader.setReader_id(rs.getString("reader_id"));
            reader.setReader_name(rs.getString("reader_name"));
            reader.setReader_special(rs.getString("reader_special"));
            reader.setReader_class(rs.getString("reader_class"));
```

```java
                reader.setReader_tel(rs.getString("reader_tel"));
                list.add(reader);
            }
        } catch (SQLException e) {
            e.printStackTrace();
        }finally{
            try {
                con.close();
            } catch (SQLException e) {
                e.printStackTrace();
            }
        }
        return list;
    }
    //删除读者
    public void deleteReader(String ID){
        int row=0;
        Connection con=DBUtil.getConnection();
        try {
            String sql="delete reader where reader_id='"+ID+"'";
            Statement sm=con.createStatement();
            row=sm.executeUpdate(sql);
        } catch (SQLException e) {
            e.printStackTrace();
        }finally{
            try {
                con.close();
            } catch (SQLException e) {
                e.printStackTrace();
            }
        }
        if(row>0){
            JOptionPane.showMessageDialog(jop,"删除成功!");
        }else{
            JOptionPane.showMessageDialog(jop,"删除失败!");
        }
    }
    //判断是否存在某个读者
    public boolean isExist(String ID){
        Connection con=DBUtil.getConnection();
        try {
            String sql="select reader_id from reader where reader_id='"+ID+"'";
            Statement sm=con.createStatement();
            ResultSet rs=sm.executeQuery(sql);
            if(rs.next()){
                return true;
```

```java
            }
        } catch (SQLException e) {
            e.printStackTrace();
        }finally{
            try {
                con.close();
            } catch (SQLException e) {
                e.printStackTrace();
            }
        }
        return false;
    }
    //查询读者类型
    public String[] seekType(){
        List<String>list=new ArrayList<String>();
        Connection con=DBUtil.getConnection();
        try {
            String sql="select * from readertype";
            Statement sm=con.createStatement();
            ResultSet rs=sm.executeQuery(sql);
            while(rs.next()){
                list.add(rs.getString("reader_type"));
            }
        } catch (SQLException e) {
            e.printStackTrace();
        }finally{
            try {
                con.close();
            } catch (SQLException e) {
                e.printStackTrace();
            }
        }
        String[] items=new String[list.size()];
        for(int i=0;i<list.size();i++){
            items[i]=list.get(i);
        }
        return items;
    }
    //查询读者类型编号
    public int seekTypeID(String s){
        int i=0;
        Connection con=DBUtil.getConnection();
        try {
            String sql="select reader_typeid from readertype where reader_type='"+s+"'";
            Statement sm=con.createStatement();
            ResultSet rs=sm.executeQuery(sql);
```

```java
            while(rs.next()){
                i=rs.getInt("reader_typeid");
            }
        } catch (SQLException e) {
            e.printStackTrace();
        }
        finally{
            try {
                con.close();
            } catch (SQLException e) {
                e.printStackTrace();
            }
        }
        return i;
    }
}
```

读者信息表模型类 ReaderTableModel：

```java
import java.util.*;
import javax.swing.table.DefaultTableModel;
import com.ytvc.zjl.entity.Readers;
public class ReaderTableModel extends DefaultTableModel{
    ReaderManage rm=new ReaderManage();
    List list=new ArrayList();
    public ReaderTableModel() {
        Object[] title={"编号","姓名","专业","班级","联系方式"};
        Object[][] date=new Object[20][5];
        super.setDataVector(date,title);
    }
    public ReaderTableModel(String ID){
        Object[] title={"编号","姓名","专业","班级","联系方式"};
        list=rm.idSeekReader(ID);
        Object[][] date=new Object[list.size()+20][5];
        for(int i=0;i<list.size();i++){
            date[i][0]=((Readers)list.get(i)).getReader_id();
            date[i][1]=((Readers)list.get(i)).getReader_name();
            date[i][2]=((Readers)list.get(i)).getReader_special();
            date[i][3]=((Readers)list.get(i)).getReader_class();
            date[i][4]=((Readers)list.get(i)).getReader_tel();
        }
        super.setDataVector(date,title);
    }
}
```

17.2.5 图书信息管理模块设计

1. 功能描述与分析

图书管理系统中主要记录图书信息,在该模块中设计图书录入信息界面,涉及出版社信息、图书类型信息等相关信息的录入与保存。图书入库时,要完成对图书的登记,应先完成出版社信息的登记,如果已有该出版社信息,就不用重复录入;图书分类信息管理,应先建立好各个类别图书信息,记录图书馆中包括图书的种类和类别,涉及的知识领域和藏书范围。

2. 涉及数据库表

完成图书信息管理模块涉及三张表,即图书类别表 booktype、出版社表 press、图书信息表 books,其表结构如表 17-6~表 17-8 所示。

表 17-6 图书类别表结构

字段名	数据类型	长度	允许空	是否主键	含义
booktype_id	int	4B	不允许	是	图书类型编号
booktype_name	varchar	20 字符	不允许		图书类型名称
booktype_memo	varchar	20 字符	允许		说明

表 17-7 出版社表结构

字段名	数据类型	长度	允许空	是否主键	含义
press_id	int	4B	不允许	是	出版社编号
press_name	varchar	50 字符	不允许		出版社名称
press_abbre	varchar	20 字符	允许		出版社简称
press_address	varchar	50 字符	允许		出版社地址
press_contact	varchar	20 字符	允许		联系人
press_tel	varchar	20 字符	允许		联系电话
email	varchar	20 字符	允许		电子邮箱

表 17-8 图书信息表结构

字段名	数据类型	长度	允许空	是否主键	含义
book_id	int	4B	不允许	是	图书编号
book_name	varchar	20 字符	不允许		图书名称
author	varchar	20 字符	不允许		作者
press_name	varchar	20 字符	允许		出版社
ISBN	varchar	20 字符	允许		图书 ISBN
publishdate	datetime	8B	允许		出版日期
pages	int	4B	允许		页数
price	float	8B	允许		价格
booktype_id	int	4B	允许		图书类型编号

续表

字段名	数据类型	长度	允许空	是否主键	含　义
stock	int	4B	允许		库存数量
brief	varchar	100 字符	允许		简介
count	int	4B	允许		借阅次数

3．界面设计分析

1）图书类型管理界面

"添加图书类型"界面如图 17-37 所示。

图 17-37　"添加图书类型"界面

2）出版社管理界面

"出版社信息设置"界面如图 17-38 所示。

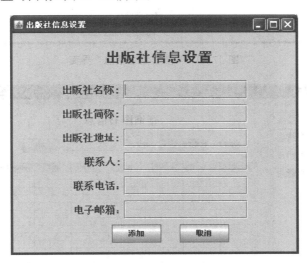

图 17-38　"出版社信息设置"界面

3）图书信息管理界面

完善图书基本信息后，对图书信息进行管理，其运行界面如图 17-39～图 17-42 所示。

图 17-39 "添加图书信息"界面

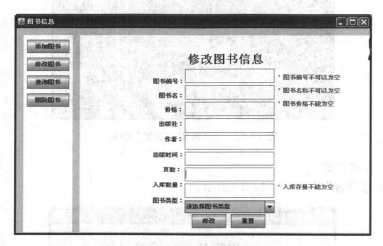

图 17-40 "修改图书信息"界面

图 17-41 "查询图书信息"界面

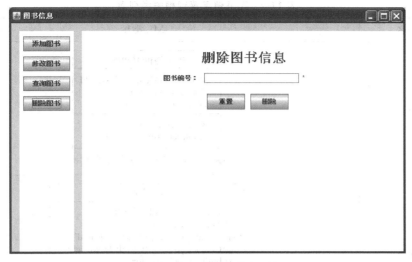

图 17-42 "删除图书信息"界面

4．分析设计相关类

图书管理界面操作中涉及的主要类及类间关系，用 UML 图描述如图 17-43 所示。

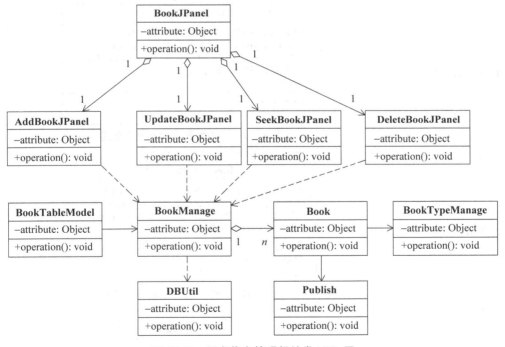

图 17-43 图书信息管理相关类 UML 图

同样，使用 DBUtil 类与数据库连接，对图书表中信息进行操作，完成对图书的管理，各个类描述如表 17-9 所示。相关类如图 17-44～图 17-53 所示。

表 17-9 图书信息管理相关类列表

类 名	UML 类图
图书类型类	**BookTypeManage** –jtfName, jtfMemo: JTextField –lName, lMemo: JLabel –submit, cancel: JButton +BookTypeManage() +actionPerformed(): void 图 17-44 "图书类型类"类图
出版社信息类	**Publish** –lName, lAbbr, lAddress: JLabel –lContact, lTel, lEmail: JLabel –jtfName, jtfAbbr, jtfAddress: JTextField –jtfContact, jtfTel, jtfEmail: JTextField –submit, cancel: JButton +publish() +actionPerformed(): void 图 17-45 "出版社信息类"类图
添加图书信息类	**AddBookJPanel** –jlBookName, jlPrice, jlPress, jlAuthor: JLabel –jlStock, jlType, jlPublishDate, jlPage: jLabel –jtfBookName, jtfPrice, jtfPress, jtfAuthor: JTextField –jtfStock, jtfPDate, jtfPage: JTextField –submit, cancel: JButton +AddBookJPanel() +actionPerformed(): void 图 17-46 "添加图书信息类"类图
修改图书信息类	**UpdateBookJPanel** -jlBookName, jlPrice, jlPress, jlAuthor: JLabel -jlStock, jlType, jlPublishDate, jlPage: JLabel -jtfBookName, jtfPrice, jtfPress, jtfAuthor: JTextField -jtfStock, jtfPDate, jtfPage: JTextField -submit, cancel: JButton +UpdateBookJPanel() +actionPerformed(): void 图 17-47 "修改图书信息类"类图
查询图书信息类	**SeekBookJPanel** –comBox: JComboBox –textfield: JTextField –submit: int +SeekBookJPanel() +actionPerformed(): void 图 17-48 "查询图书信息类"类图

续表

类　　名	UML 类图
删除图书信息类	**DeleteBookJPanel** −jtfID: JTextField −submit: JButton +DeleteBookJPanel() +actionPerformed(): void 图 17-49　"删除图书信息类"类图
图书信息管理类	**BookJPanel** −AddBookJPanel: JPanel −UpdateBookJPanel: JPanel −DeleteBookJPanel: JPanel −SeekBookJPanel: JPanel −add, update, delete, seek: JButton +BookJPanel() +actionPerformed(): void 图 17-50　"图书信息管理类"类图
图书实体类	**Book** −book_id: String −book_name: String −price: double −press: String −author: String −stock: int −type: String −count: int +get×××(): String +get×××(): int +set×××(): void 图 17-51　"图书实体类"类图
图书信息操作类	**BookManage** +borrowPaihang(): Object +addBook(book: Book): void +seekType(): Object[] +idSeekBook(id: String): Book +updateBook(Book book: int): void +seekBook(s: String, index: int): Object[][] +deleteBook(ID: String): void +typeCount(): List 图 17-52　"图书信息操作类"类图
图书信息表模型类	**BookTableModel** +BookTableModel() +BookTableModel(b: boolean) +BookTableModel(s: String, index: int) 图 17-53　"图书信息表模型类"类图

5. 相关类主要代码

该模块设计涉及类的操作与读者的操作十分相似,大家可参考读者信息管理模块进行操作,也可查看附带程序代码。

17.2.6 图书借阅/归还操作模块设计

1. 功能描述与分析

该模块实现图书馆图书借阅操作,是图书管理系统中的核心功能,实现读者对图书的借入操作,记录着读者借阅图书的信息、借阅时间,同时能查询图书馆中借阅图书排行榜;实现读者对图书的归还操作,记录着读者对图书归还的信息,返还时间,是否需交罚款和应交罚款金额。

2. 建立数据库表

图书借阅/归还模块涉及图书信息表 books、读者信息表 reader 和图书借阅信息表 borrowinfo,其中图书借阅信息表结构如表 17-10 所示。

表 17-10 图书借阅信息表结构

字段名	数据类型	长度	允许空	是否主键	含义
borrow_id	int	4B	不允许	是	借阅编号
reader_id	int	4B	允许		读者编号
book_id	int	4B	允许		图书编号
starttime	datetime	8B	允许		借阅时间
stoptime	datetime	8B	允许		归还时间
penalty	int	4B	允许		罚款金额

3. 界面设计分析

图书管理员使用图书借还管理模块输入读者编号和图书编号,可以查询是否存在该用户和图书,实现图书借阅,并能实现借阅次数查询,如图 17-54 所示。

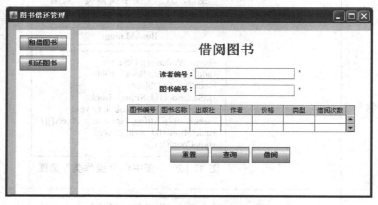

图 17-54 "图书借阅操作"界面

在读者归还操作界面中输入相应信息,可以查询用户是否借阅某本图书,如果没有该用户借阅该图书的信息,提示用户没有借阅信息;如果有借阅信息,可以完成图书的归还,如图 17-55~图 17-57 所示。

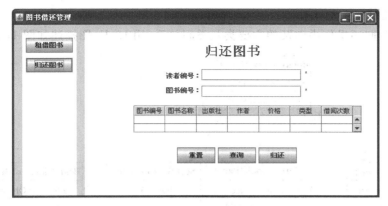

图 17-55 "图书归还操作"界面

图 17-56 归还失败提示界面

图 17-57 归还成功提示界面

以汇总方式指定查询日期,查询某段时间借阅图书信息及相关是否罚款信息,如图 17-58 和图 17-59 所示。

图 17-58 "日期选择"界面

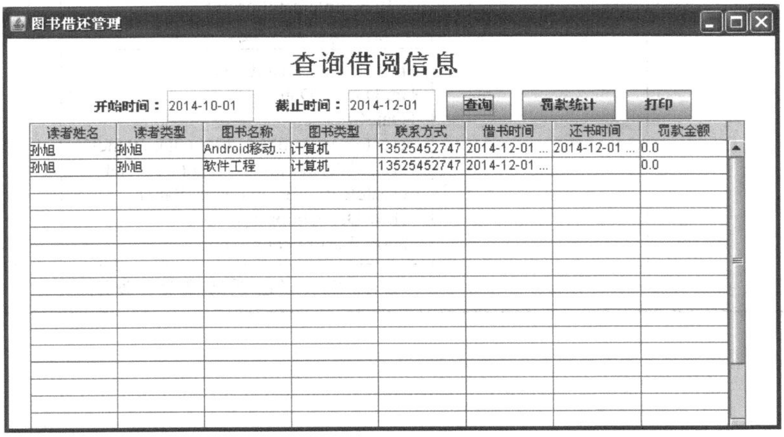

图 17-59 "查询借阅信息"界面

4. 分析设计相关类

图书借阅/归还操作中涉及的主要类及类间关系,用 UML 图描述如图 17-60 所示。

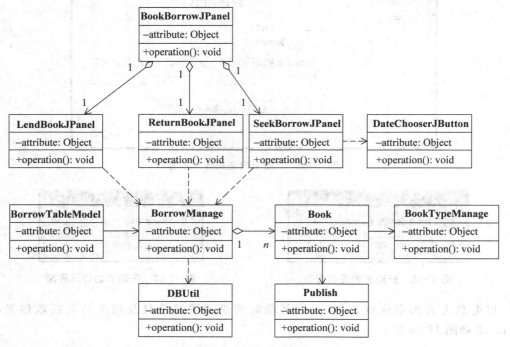

图 17-60 图书借阅/归还操作相关类 UML 图

使用 DBUtil 类与数据库连接,完成读者对图书借阅、归还操作,实现图书借还管理,主要类描述如表 17-11 所示。相关类图如图 17-61~图 17-67 所示。

表 17-11 图书借阅/归还操作相关类列表

类 名	UML 图
日期选择类	**DateChooserJButton** −dateChooser: DateChooser −sdf: SimpleDateFormat −originalText: String +DateChooserJButton() +getDate(): Date +DateChooser() +createYearAndMonthPanal(): JPanel +createWeekAndDayPanal(): JPanel +createButtonBarPanel(): JPanel +showDateChooser(): void +actionPerfomed(): void 图 17-61 "日期选择类" 类图

续表

类　　名	UML 图
图书管理操作类	**BorrowManage** +borrowBook(): void +returnBook(): void +isBorrow(): boolean +seekBorrow(): Object[][] 图 17-62　"图书管理操作类"类图
图书借阅操作类	**LendBookJPanel** −jtfReaderID, jtfBookID: JTextField −jbSeek, jbReset, jbSubmit: JButton −bm: BorrowManage +LendBookJPanel() +actionPerformed(): void 图 17-63　"图书借阅操作类"类图
图书归还操作类	**ReturnBookJPanel** −jtfReaderID, jtfBookID: JTextField −jbSeek, jbReset, jbSubmit: JButton −bm: BorrowManage +ReturnBookJPanel() +actionPerformed(): void 图 17-64　"图书归还操作类"类图
图书借阅/归还操作类	**BookBorrowJPanel** −jpLeft: JPanel −lendBook, returnBook: JButton +BookBorrowJPanel() +actionPerformed(): void 图 17-65　"图书借阅/归还操作类"类图
图书查询操作类	**SeekBorrowJPanel** −dcjStart, dcjStop: DateChooserJButton −jt: JTable −jbSeek: JButton +SeekBorrowJPanel() +actionPerformed(): void 图 17-66　"图书查询操作类"类图
图书信息表模型类	**BorrowTableModel** −bm: BorrowManage +BorrowTableModel() +BorrowTableModel(start: String, stop: String) +BorrowTableModel(start: String, stop: String, a: int) +BorrowTableModel(a: int) 图 17-67　"图书信息表模型类"类图

5. 相关类主要代码

该模块设计涉及类的主要代码如下,详见附带代码。

图书借阅/归还操作类 BookBorrowJPanel：

```java
public class BookBorrowJPanel extends JPanel implements ActionListener{
    JPanel jpLeft;
    JButton lendBook,returnBook;
    public static LendBookJPanel lbj=new LendBookJPanel();
    public static ReturnBookJPanel rbj=new ReturnBookJPanel();
    public BookBorrowJPanel() {
        lbj.setVisible(false);
        rbj.setVisible(false);
        setLayout(null);
        jpLeft=new JPanel();
        lendBook=new JButton("租借图书");
        lendBook.addActionListener(this);
        returnBook=new JButton("归还图书");
        returnBook.addActionListener(this);
        jpLeft.add(Box.createHorizontalStrut(10));
        jpLeft.add(lendBook);
        jpLeft.add(Box.createHorizontalStrut(10));
        jpLeft.add(Box.createHorizontalStrut(10));
        jpLeft.add(Box.createHorizontalStrut(10));
        jpLeft.add(returnBook);
        jpLeft.setBounds(15,15,100,370);
        jpLeft.setBackground(Color.white);
        add(jpLeft);
        lbj.setBounds(130,15,685,370);
        add(lbj);
        rbj.setBounds(130,15,685,370);
        add(rbj);
    }
    public void actionPerformed(ActionEvent e) {
        if(e.getSource()==lendBook){
            MainInterface.jp.setVisible(false);
            rbj.setVisible(false);
            lbj.setVisible(true);
            lbj.init();
        }
        if(e.getSource()==returnBook){
            MainInterface.jp.setVisible(false);
            lbj.setVisible(false);
            rbj.setVisible(true);
            rbj.init();
        }
```

        }
}

图书借阅操作类 LendBookJPanel：

```
public class LendBookJPanel extends JPanel implements ActionListener,
FocusListener{
    Box box1,box2,box3,boxBase;
    JLabel jl,jlReaderID,jlBookID,jl1,jl2;
    JTextField jtfReaderID,jtfBookID;
    JButton jbSeek,jbReset,jbSubmit;
    JScrollPane js;
    JTable jt;
    ReaderManage rm=new ReaderManage();
    BookManage bkm=new BookManage();
    BorrowManage bm=new BorrowManage();
    public LendBookJPanel() {
        jl=new JLabel("借阅图书");
        jl.setFont(new Font("宋体",Font.BOLD,25));
        jl.setForeground(Color.blue);
        jl.setBounds(220,10,200,50);
        add(jl);
        jlReaderID=new JLabel("读者编号：");
        jlBookID=new JLabel("图书编号：");
        jtfReaderID=new JTextField();
        jtfBookID=new JTextField();
        jtfReaderID.addFocusListener(this);
        jtfBookID.addFocusListener(this);
        jl1=new JLabel(" * ");
        jl1.setForeground(Color.red);
        jl2=new JLabel(" * ");
        jl2.setForeground(Color.red);
        box1=Box.createVerticalBox();
        box1.add(jlReaderID);
        box1.add(Box.createVerticalStrut(10));
        box1.add(jlBookID);
        box2=Box.createVerticalBox();
        box2.add(jtfReaderID);
        box2.add(Box.createVerticalStrut(10));
        box2.add(jtfBookID);
        boxBase=Box.createHorizontalBox();
        boxBase.add(box1);
        boxBase.add(box2);
        boxBase.setBounds(150,70,250,55);
        add(boxBase);
        box3=Box.createVerticalBox();
```

```java
            box3.add(jl1);
            box3.add(Box.createVerticalStrut(15));
            box3.add(jl2);
            box3.setBounds(405,70,150,55);
            add(box3);
            jt=new JTable(new BookTableModel(true));
            js=new JScrollPane(jt);
            js.setBounds(90,140,420,52);
            add(js);
            jbReset=new JButton("重置");
            jbReset.setBounds(170,220,70,30);
            jbSeek=new JButton("查询");
            jbSeek.setBounds(245,220,70,30);
            jbSubmit=new JButton("借阅");
            jbSubmit.setBounds(320,220,70,30);
            jbReset.addActionListener(this);
            jbSeek.addActionListener(this);
            jbSubmit.addActionListener(this);
            add(jbReset);
            add(jbSeek);
            add(jbSubmit);
            setBackground(Color.white);
            setLayout(null);
        }
        public void actionPerformed(ActionEvent e) {
            if(e.getSource()==jbReset){
                init();
            }
            if(e.getSource()==jbSeek){
                if(jtfBookID.getText().equals("")){
                }else{
                    jt.setModel(new BookTableModel(jtfBookID.getText(),0));
                }
            }
            if(e.getSource()==jbSubmit){
            if(jtfBookID.getText().equals("")||jtfReaderID.getText().equals("")){
                }else{
                    bm.borrowBook(jtfBookID.getText(),jtfReaderID.getText());
                    init();
                }
            }
        }
        public void focusGained(FocusEvent e) {}
        public void focusLost(FocusEvent e) {
            if(e.getSource()==jtfReaderID){
```

```java
            if(jtfReaderID.getText().equals("")){
                jl1.setText("* 读者编号不能为空");
            }else{
                if(rm.isExist(jtfReaderID.getText())){
                    jl1.setText("* ");
                }else{
                    jl1.setText("* 没有此读者信息");
                }
            }
        }
        if(e.getSource()==jtfBookID){
            Books book;
            if(jtfBookID.getText().equals("")){
                jl2.setText("* 图书编号不能为空");
            }else{
                book=bkm.idSeekBook(jtfBookID.getText());
                if(book==null){
                    jl2.setText("* 没有此图书的信息");
                }else{
                    jl2.setText("* ");
                    jt.setModel(new BookTableModel(jtfBookID.getText(),0));
                }
            }
        }
    }
    public void init(){
        jtfReaderID.setText("");
        jtfBookID.setText("");
        jt.setModel(new BookTableModel(true));
        jl1.setText("* ");
        jl2.setText("* ");
    }
}
```

图书归还操作类 ReturnBookJPanel：

```java
public class ReturnBookJPanel extends JPanel implements ActionListener,
FocusListener{
    Box box1,box2,box3,boxBase;
    JLabel jl,jlReaderID,jlBookID,jl1,jl2;
    JTextField jtfReaderID,jtfBookID;
    JButton jbSeek,jbReset,jbSubmit;
    JScrollPane js;
    JTable jt;
    JOptionPane jop=new JOptionPane();
    ReaderManage rm=new ReaderManage();
```

```java
BookManage bkm=new BookManage();
BorrowManage bm=new BorrowManage();
public ReturnBookJPanel() {
    jl=new JLabel("归还图书");
    jl.setFont(new Font("宋体",Font.BOLD,25));
    jl.setForeground(Color.blue);
    jl.setBounds(220,10,200,50);
    add(jl);
    jlReaderID=new JLabel("读者编号：");
    jlBookID=new JLabel("图书编号：");
    jtfReaderID=new JTextField();
    jtfBookID=new JTextField();
    jtfReaderID.addFocusListener(this);
    jtfBookID.addFocusListener(this);
    jl1=new JLabel(" * ");
    jl1.setForeground(Color.red);
    jl2=new JLabel(" * ");
    jl2.setForeground(Color.red);
    box1=Box.createVerticalBox();
    box1.add(jlReaderID);
    box1.add(Box.createVerticalStrut(10));
    box1.add(jlBookID);
    box2=Box.createVerticalBox();
    box2.add(jtfReaderID);
    box2.add(Box.createVerticalStrut(10));
    box2.add(jtfBookID);
    boxBase=Box.createHorizontalBox();
    boxBase.add(box1);
    boxBase.add(box2);
    boxBase.setBounds(150,70,250,55);
    add(boxBase);
    box3=Box.createVerticalBox();
    box3.add(jl1);
    box3.add(Box.createVerticalStrut(15));
    box3.add(jl2);
    box3.setBounds(405,70,150,55);
    add(box3);
    jt=new JTable(new BookTableModel(true));
    js=new JScrollPane(jt);
    js.setBounds(90,140,420,52);
    add(js);
    jbReset=new JButton("重置");
    jbReset.setBounds(170,220,70,30);
    jbSeek=new JButton("查询");
    jbSeek.setBounds(245,220,70,30);
```

```java
        jbSubmit=new JButton("归还");
        jbSubmit.setBounds(320,220,70,30);
        jbReset.addActionListener(this);
        jbSeek.addActionListener(this);
        jbSubmit.addActionListener(this);
        add(jbReset);
        add(jbSeek);
        add(jbSubmit);
        setBackground(Color.white);
        setLayout(null);
    }
    public void actionPerformed(ActionEvent e) {
        if(e.getSource()==jbReset){
            init();
        }
        if(e.getSource()==jbSeek){
if(jtfBookID.getText().equals("")||jtfReaderID.getText().equals("")){}
            else{
                if(!(bm.isBorrow(jtfReaderID.getText(),jtfBookID.getText()))){
            JOptionPane.showMessageDialog(jop,"没有此图书的借阅信息!");
                }else{
            jt.setModel(new BookTableModel(jtfBookID.getText(),0));
                }
            }
        }
        if(e.getSource()==jbSubmit){
if(jtfBookID.getText().equals("")||jtfReaderID.getText().equals("")){}
            else{
                if(!(bm.isBorrow(jtfReaderID.getText(),jtfBookID.getText()))){
                JOptionPane.showMessageDialog(jop,"没有此图书的借阅信息!");
                }else{
                    bm.returnBook(jtfBookID.getText(),jtfReaderID.getText());
                    init();
                }
            }
        }
    }
    public void focusGained(FocusEvent e) {}
    public void focusLost(FocusEvent e) {
        if(e.getSource()==jtfReaderID){
            if(jtfReaderID.getText().equals("")){
                jl1.setText(" * 读者编号不能为空");
            }else{
                if(rm.isExist(jtfReaderID.getText())){
                    jl1.setText(" * ");
```

```java
            }else{
                jl1.setText("* 没有此读者信息");
            }
        }
    }
    if(e.getSource()==jtfBookID){
        Books book;
        if(jtfBookID.getText().equals("")){
            jl2.setText("* 图书编号不能为空");
        }else{
            book=bkm.idSeekBook(jtfBookID.getText());
            if(book==null){
                jl2.setText("* 没有此图书的信息");
            }else{
                jl2.setText("* ");
                jt.setModel(new BookTableModel(jtfBookID.getText(),0));
            }
        }
    }
}
public void init(){
    jtfReaderID.setText("");
    jtfBookID.setText("");
    jt.setModel(new BookTableModel(true));
    jl1.setText("* ");
    jl2.setText("* ");
}
```

图书查询操作类 SeekBorrowJPanel：

```java
public class SeekBorrowJPanel extends JPanel implements ActionListener{
    JLabel jl,jl1,jl2;
    DateChooserJButton dcjStart,dcjStop;
    JScrollPane js;
    JTable jt;
    Box box;
    JButton jbSeek,jb,dayin;
    BorrowManage bm=new BorrowManage();
    public SeekBorrowJPanel() {
        setLayout(null);
        jl=new JLabel("查询借阅信息");
        jl.setFont(new Font("宋体",Font.BOLD,25));
        jl.setForeground(Color.blue);
        jl.setBounds(260,10,200,30);
        add(jl);
```

```java
        jl1=new JLabel("开始时间：");
        jl2=new JLabel("截止时间：");
        dcjStart=new DateChooserJButton();
        dcjStop=new DateChooserJButton();
        dcjStart.setBorder(BorderFactory.createLineBorder(new Color(51,255,51)));
        dcjStop.setBorder(BorderFactory.createLineBorder(new Color(51,255,51)));
        jbSeek=new JButton("查询");
        jbSeek.addActionListener(this);
        jb=new JButton("罚款统计");
        jb.addActionListener(this);
        dayin=new JButton("打印");
        dayin.addActionListener(this);
        box=Box.createHorizontalBox();
        box.add(jl1);
        box.add(Box.createHorizontalStrut(2));
        box.add(dcjStart);
        box.add(Box.createHorizontalStrut(20));
        box.add(jl2);
        box.add(Box.createHorizontalStrut(2));
        box.add(dcjStop);
        box.add(Box.createHorizontalStrut(10));
        box.add(jbSeek);
        box.add(Box.createHorizontalStrut(10));
        box.add(jb);
        box.add(Box.createHorizontalStrut(10));
        box.add(dayin);
        box.setBounds(80,50,550,30);
        add(box);
        jt=new JTable(new BorrowTableModel());
        js=new JScrollPane(jt);
        js.setBounds(20,80,660,300);
        add(js);
        setBackground(Color.white);
    }
    public void actionPerformed(ActionEvent e) {
        if(e.getSource()==jbSeek){
            jt.setModel(new BorrowTableModel(dcjStart.getText(),dcjStop.getText()));
        }
        if(e.getSource()==jb){
            new MoneySeek();
        }
        if(e.getSource()==dayin){
            bm.borrowReport();
        }
    }
```

```java
    public void init(){
        jt.setModel(new BorrowTableModel());
    }
}
```

图书管理操作类 BorrowManage：

```java
public class BorrowManage {
    JOptionPane jop=new JOptionPane();
    //借阅图书
    public void borrowBook(String bookID,String readerID){
        int row=0;
        int stock=0;
        SimpleDateFormat sdf=new SimpleDateFormat("yyyy-MM-dd");
        Date d=new Date();
        String time=sdf.format(d);
        Connection con=DBUtil.getConnection();
        try {
            Statement sm=con.createStatement();
            String sql="select * from books where book_id="+bookID;
            ResultSet rs=sm.executeQuery(sql);
            if(rs.next()){
                stock=rs.getInt("stock");
            }
            sql="select * from borrowinfo where reader_id='"+readerID+"' and
                book_id='"+bookID+"' and stopTime is null";
            rs=sm.executeQuery(sql);
            if(rs.next()){
                JOptionPane.showMessageDialog(jop,"你已经借阅过此图书!");
            }else{
                if(stock>0){
                    con.setAutoCommit(false);
            sql="insert into borrowInfo values('"+readerID+"','"+bookID+"',
                '"+time+"',null,0.0)";
                    row=sm.executeUpdate(sql);
            sql="update books set stock=stock-1,count=count+1 where book_id="
                +bookID;
                    row=sm.executeUpdate(sql)+row;
                    con.commit();
                    con.setAutoCommit(true);
                }else{
                    JOptionPane.showMessageDialog(jop,"此图书暂缺!");
                }
            }
        } catch (SQLException e) {
            e.printStackTrace();
```

```java
        }finally{
            try {
                con.close();
            } catch (SQLException e) {
                e.printStackTrace();
            }
        }
        if(row>1){
            JOptionPane.showMessageDialog(jop,"借阅成功!");
        }else{
            JOptionPane.showMessageDialog(jop,"借阅失败!");
        }
    }
    //归还图书
    public void returnBook(String bookID,String readerID){
        int row=0;
        int day=0;
        String type=null;
        double penalty=0;
        SimpleDateFormat sdf=new SimpleDateFormat("yyyy-MM-dd");
        Date d=new Date();
        String time=sdf.format(d);
        try {
            Connection con=DBUtil.getConnection();
            String startTime=null;
            String sql="select * from borrowInfo where reader_id="+readerID+
                    " and book_id="+bookID+" and stopTime is null";
            Statement sm=con.createStatement();
            ResultSet rs=sm.executeQuery(sql);
            if(rs.next()){
                    sql="update books set stock=stock+1 where book_id="+bookID;
                    row=sm.executeUpdate(sql)+row;
                    con.commit();
                    con.setAutoCommit(true);
            }
            }else{
JOptionPane.showMessageDialog(jop,"你没有借阅此图书,或者已经归还了!");
            }
            con.close();
        } catch (SQLException e) {
                e.printStackTrace();
        }
        if(row>1){
            JOptionPane.showMessageDialog(jop,"归还成功!");
        }else{
```

```java
            JOptionPane.showMessageDialog(jop,"归还失败!");
        }
    }
    //判断是否有此读者的借阅信息
    public boolean isBorrow(String readerID,String bookID){
        Connection con=DBUtil.getConnection();
        try {
            String sql="select * from borrowInfo where reader_id="+readerID+
                    " and book_id="+bookID+"";
            Statement sm=con.createStatement();
            ResultSet rs=sm.executeQuery(sql);
            if(rs.next()){
                return true;
            }
        } catch (SQLException e) {
            e.printStackTrace();
        }finally{
            try {
                con.close();
            } catch (SQLException e) {
                e.printStackTrace();
            }
        }
        return false;
    }
    //根据开始和结束时间段查询图书
    public Object[][] seekBorrow(String start,String stop){
        Borrow borrow;
        List<Borrow>list=new ArrayList<Borrow>();
        Connection con=DBUtil.getConnection();
        try {
            String sql="select * from books,reader,borrowInfo where books.book_
                    id=borrowInfo.book_id and reader.reader_id=borrowInfo.
                    reader_id and startTime between '"+start+"' and '"+stop+
                    "'";
            Statement sm=con.createStatement();
            ResultSet rs=sm.executeQuery(sql);
            while(rs.next()){
                borrow=new Borrow();
                borrow.setReader_name(rs.getString("reader_name"));
                borrow.setReader_type(rs.getString(11));
                borrow.setBook_name(rs.getString("book_name"));
                borrow.setBook_type(rs.getString(7));
                borrow.setPhone(rs.getString("phone"));
                borrow.setStartTime(rs.getString("startTime"));
```

```java
                borrow.setStopTime(rs.getString("stopTime"));
                borrow.setPenalty(rs.getDouble("penalty"));
                list.add(borrow);
            }
        } catch (SQLException e) {
            e.printStackTrace();
        }finally{
            try {
                con.close();
            } catch (SQLException e) {
                e.printStackTrace();
            }
        }
        Object[][] date=new Object[list.size()+20][8];
        for(int i=0;i<list.size();i++){
            date[i][0]=list.get(i).getReader_name();
            date[i][1]=list.get(i).getReader_type();
            date[i][2]=list.get(i).getBook_name();
            date[i][3]=list.get(i).getBook_type();
            date[i][4]=list.get(i).getPhone();
            date[i][5]=list.get(i).getStartTime();
            date[i][6]=list.get(i).getStopTime();
            date[i][7]=list.get(i).getPenalty();
        }
        return date;
    }
    //查询所有借阅信息
    public Object[][] seekBorrow(){
        Borrow borrow;
        List<Borrow>list=new ArrayList<Borrow>();
        Connection con=DBUtil.getConnection();
        try {
            String sql="select * from books,reader,borrowInfo where books.book_
                    id=borrowInfo.book_id and reader.reader_id=borrowInfo.
                    reader_id";
            Statement sm=con.createStatement();
            ResultSet rs=sm.executeQuery(sql);
            while(rs.next()){
                borrow=new Borrow();
                borrow.setReader_name(rs.getString("reader_name"));
                borrow.setReader_type(rs.getString(12));
                borrow.setBook_name(rs.getString("book_name"));
                borrow.setBook_type(rs.getString(7));
                borrow.setPhone(rs.getString("phone"));
                borrow.setStartTime(rs.getString("startTime"));
```

```java
            borrow.setStopTime(rs.getString("stopTime"));
            borrow.setPenalty(rs.getDouble("penalty"));
            list.add(borrow);
        }
    } catch (SQLException e) {
        e.printStackTrace();
    }finally{
        try {
            con.close();
        } catch (SQLException e) {
            e.printStackTrace();
        }
    }
    Object[][] date=new Object[list.size()+20][8];
    for(int i=0;i<list.size();i++){
        date[i][0]=list.get(i).getReader_name();
        date[i][1]=list.get(i).getReader_type();
        date[i][2]=list.get(i).getBook_name();
        date[i][3]=list.get(i).getBook_type();
        date[i][4]=list.get(i).getPhone();
        date[i][5]=list.get(i).getStartTime();
        date[i][6]=list.get(i).getStopTime();
        date[i][7]=list.get(i).getPenalty();
    }
    return date;
}
//查询所有罚款的借阅信息
public Object[][] seekBorrow(int x){
    Borrow borrow;
    List<Borrow> list=new ArrayList<Borrow>();
    Connection con=DBUtil.getConnection();
    try {
        String sql="select * from books,reader,borrowInfo where books.book_
                    id=borrowInfo.book_id and reader.reader_id=borrowInfo.
                    reader_id and penalty>0";
        Statement sm=con.createStatement();
        ResultSet rs=sm.executeQuery(sql);
        while(rs.next()){
            borrow=new Borrow();
            borrow.setReader_name(rs.getString("reader_name"));
            borrow.setReader_type(rs.getString(11));
            borrow.setBook_name(rs.getString("book_name"));
            borrow.setBook_type(rs.getString(7));
            borrow.setPhone(rs.getString("phone"));
            borrow.setStartTime(rs.getString("startTime"));
```

```java
                    borrow.setStopTime(rs.getString("stopTime"));
                    borrow.setPenalty(rs.getDouble("penalty"));
                    list.add(borrow);
            }
        } catch (SQLException e) {
                e.printStackTrace();
        }finally{
            try {
                con.close();
            } catch (SQLException e) {
                e.printStackTrace();
            }
        }
        Object[][] date=new Object[list.size()+20][8];
        for(int i=0;i<list.size();i++){
            date[i][0]=list.get(i).getReader_name();
            date[i][1]=list.get(i).getReader_type();
            date[i][2]=list.get(i).getBook_name();
            date[i][3]=list.get(i).getBook_type();
            date[i][4]=list.get(i).getPhone();
            date[i][5]=list.get(i).getStartTime();
            date[i][6]=list.get(i).getStopTime();
            date[i][7]=list.get(i).getPenalty();
        }
        return date;
    }
//按时间段查询罚款的借阅信息
public Object[][] seekBorrow(String start,String stop,int x){
    Borrow borrow;
    List<Borrow>list=new ArrayList<Borrow>();
    Connection con=DBUtil.getConnection();
    try {
            String sql="select * from books,reader,borrowInfo where books.book_
                    id=borrowInfo.book_id and reader.reader_id=borrowInfo.
                    reader_id and startTime between '"+start+"' and '"+stop+
                    "' and penalty>0";
            Statement sm=con.createStatement();
            ResultSet rs=sm.executeQuery(sql);
            while(rs.next()){
                borrow=new Borrow();
                borrow.setReader_name(rs.getString("reader_name"));
                borrow.setReader_type(rs.getString(11));
                borrow.setBook_name(rs.getString("book_name"));
                borrow.setBook_type(rs.getString(7));
                borrow.setPhone(rs.getString("phone"));
```

```java
                borrow.setStartTime(rs.getString("startTime"));
                borrow.setStopTime(rs.getString("stopTime"));
                borrow.setPenalty(rs.getDouble("penalty"));
                list.add(borrow);
            }
        } catch (SQLException e) {
            e.printStackTrace();
        }finally{
            try {
                con.close();
            } catch (SQLException e) {
                e.printStackTrace();
            }
        }
        Object[][] date=new Object[list.size()+20][8];
        for(int i=0;i<list.size();i++){
            date[i][0]=list.get(i).getReader_name();
            date[i][1]=list.get(i).getReader_type();
            date[i][2]=list.get(i).getBook_name();
            date[i][3]=list.get(i).getBook_type();
            date[i][4]=list.get(i).getPhone();
            date[i][5]=list.get(i).getStartTime();
            date[i][6]=list.get(i).getStopTime();
            date[i][7]=list.get(i).getPenalty();
        }
        return date;
    }
    //统计一共罚款数
    public double sumMoney(){
        double sum=0;
        Connection con=DBUtil.getConnection();
        try {
            String sql="select penalty from borrowInfo";
            Statement sm=con.createStatement();
            ResultSet rs=sm.executeQuery(sql);
            while(rs.next()){
                sum+=rs.getDouble(1);
            }
        } catch (SQLException e) {
            e.printStackTrace();
        }finally{
            try {
                con.close();
            } catch (SQLException e) {
                e.printStackTrace();
```

```
                }
            }
            return sum;
        }
    }
```

图书信息表模型类 BorrowTableModel：

```
public class BorrowTableModel extends DefaultTableModel {
    BorrowManage bm=new BorrowManage();
    public BorrowTableModel() {
        Object[] title={"读者姓名","读者类型","图书名称","图书类型","联系方式","借
                        书时间","还书时间","罚款金额"};
        Object[][] date=bm.seekBorrow();
        super.setDataVector(date,title);
    }
    public BorrowTableModel(String start,String stop) {
        Object[] title={"读者姓名","读者类型","图书名称","图书类型","联系方式","借
                        书时间","还书时间","罚款金额"};
        Object[][] date=bm.seekBorrow(start,stop);
        super.setDataVector(date,title);
    }
    public BorrowTableModel(String start,String stop,int a){
        Object[] title={"读者姓名","读者类型","图书名称","图书类型","联系方式","借
                        书时间","还书时间","罚款金额"};
        Object[][] date=bm.seekBorrow(start,stop,1);
        super.setDataVector(date,title);
    }
    public BorrowTableModel(int a){
        Object[] title={"读者姓名","读者类型","图书名称","图书类型","联系方式","借
                        书时间","还书时间","罚款金额"};
        Object[][] date=bm.seekBorrow(1);
        super.setDataVector(date,title);
    }
}
```

日期选择类 DateChooserJButton：

```
public class DateChooserJButton extends JTextField {
    private DateChooser dateChooser=null;
    private String preLabel="";
    private String originalText=null;
    private SimpleDateFormat sdf=null;
    public DateChooserJButton() {
        this(getNowDate());
    }
    public DateChooserJButton(String dateString) {
```

```java
        this();
        setText(getDefaultDateFormat(),dateString);
        //保存原始日期时间
        initOriginalText(dateString);
    }
    public DateChooserJButton(SimpleDateFormat df,String dateString) {
        this();
        setText(df,dateString);
        //记忆当前的日期格式化器
        this.sdf=df;
        //记忆原始日期时间
        Date originalDate=null;
        try {
            originalDate=df.parse(dateString);
        } catch (ParseException ex) {
            originalDate=getNowDate();
        }
        initOriginalText(originalDate);
    }
    public DateChooserJButton(Date date) {
        this("",date);
        //记忆原始日期时间
        initOriginalText(date);
    }
    public DateChooserJButton(String preLabel,Date date) {
        if (preLabel!=null) {
            this.preLabel=preLabel;
        }
        setDate(date);
        //记忆原始日期时间
        initOriginalText(date);
        setBorder(null);
        setCursor(new Cursor(Cursor.HAND_CURSOR));
        addMouseListener(new MouseAdapter() {
            public void mouseClicked(MouseEvent e) {
                if (dateChooser==null) {
                    dateChooser=new DateChooser();
                }
                Point p=getLocationOnScreen();
                p.y=p.y+30;
                dateChooser.showDateChooser(p);
            }
        });
    }
    private static Date getNowDate() {
```

```java
        return Calendar.getInstance().getTime();
    }
    private static SimpleDateFormat getDefaultDateFormat() {
        return new SimpleDateFormat("yyyy-MM-dd");
    }
    /**
     * 得到当前使用的日期格式化器
     */
    public SimpleDateFormat getCurrentSimpleDateFormat() {
        if (this.sdf!=null) {
            return sdf;
        } else {
            return getDefaultDateFormat();
        }
    }
    //保存原始日期时间
    private void initOriginalText(String dateString) {
        this.originalText=dateString;
    }
    //保存原始日期时间
    private void initOriginalText(Date date) {
        this.originalText=preLabel+getDefaultDateFormat().format(date);
    }
    /**
     * 得到当前记忆的原始日期时间
     */
    public String getOriginalText() {
        return originalText;
    }
    //覆盖父类的方法
    public void setText(String s) {
        Date date;
        try {
            date=getDefaultDateFormat().parse(s);
        } catch (ParseException e) {
            date=getNowDate();
        }
        setDate(date);
    }
    public void setText(SimpleDateFormat df,String s) {
        Date date;
        try {
            date=df.parse(s);
        } catch (ParseException e) {
            date=getNowDate();
```

```java
        }
        setDate(date);
    }
    public void setDate(Date date) {
        super.setText(preLabel+getDefaultDateFormat().format(date));
    }
    public Date getDate() {
        String dateString=getText().substring(preLabel.length());
        try {
            SimpleDateFormat currentSdf=getCurrentSimpleDateFormat();
            return currentSdf.parse(dateString);
        } catch (ParseException e) {
            return getNowDate();
        }
    }
    public void addActionListener(ActionListener listener) {
    }
    /**
     * 内部类,主要是定义一个JPanel,然后把日历相关的所有内容填入本JPanel,
     * 然后再创建一个JDialog,把本内部类定义的JPanel放入JDialog的内容区
     */
    private class DateChooser extends JPanel implements ActionListener,
            ChangeListener {
        int startYear=1980;                          //默认最小显示年份
        int lastYear=2050;                           //默认最大显示年份
        int width=390;                               //界面宽度
        int height=210;                              //界面高度
        Color backGroundColor=Color.gray;            //底色
        //月历表格配色----------------//
        Color palletTableColor=Color.white;          //日历表底色
        Color todayBackColor=Color.orange;           //今天背景色
        Color weekFontColor=Color.blue;              //星期文字色
        Color dateFontColor=Color.black;             //日期文字色
        Color weekendFontColor=Color.red;            //周末文字色
        //控制条配色-------------------//
        Color controlLineColor=Color.pink;           //控制条底色
        Color controlTextColor=Color.white;          //控制条标签文字色
        Color rbFontColor=Color.white;               //RoundBox文字色
        Color rbBorderColor=Color.red;               //RoundBox边框色
        Color rbButtonColor=Color.pink;              //RoundBox按钮色
        Color rbBtFontColor=Color.red;               //RoundBox按钮文字色
        /**单击DateChooserButton时弹出的对话框,日历内容在这个对话框内 */
        JDialog dialog;
        JSpinner yearSpin;
        JSpinner monthSpin;
```

```java
    JSpinner daySpin;
    JSpinner hourSpin;
    JSpinner minuteSpin;
    JSpinner secondSpin;
    JButton[][] daysButton=new JButton[6][7];
    DateChooser() {
        setLayout(new BorderLayout());
        setBorder(new LineBorder(backGroundColor,2));
        setBackground(backGroundColor);
        JPanel topYearAndMonth=createYearAndMonthPanal();
        add(topYearAndMonth,BorderLayout.NORTH);
        JPanel centerWeekAndDay=createWeekAndDayPanal();
        add(centerWeekAndDay,BorderLayout.CENTER);
        JPanel buttonBarPanel=createButtonBarPanel();
        this.add(buttonBarPanel,java.awt.BorderLayout.SOUTH);
    }
    private JPanel createYearAndMonthPanal() {
        Calendar c=getCalendar();
        int currentYear=c.get(Calendar.YEAR);
        int currentMonth=c.get(Calendar.MONTH)+1;
        int currentHour=c.get(Calendar.HOUR_OF_DAY);
        int currentMinute=c.get(Calendar.MINUTE);
        int currentSecond=c.get(Calendar.SECOND);
        JPanel result=new JPanel();
        result.setLayout(new FlowLayout());
        result.setBackground(controlLineColor);
        yearSpin=new JSpinner(new SpinnerNumberModel(currentYear,
                startYear,lastYear,1));
        yearSpin.setPreferredSize(new Dimension(48,20));
        yearSpin.setName("Year");
        yearSpin.setEditor(new JSpinner.NumberEditor(yearSpin,"####"));
        yearSpin.addChangeListener(this);
        result.add(yearSpin);
        JLabel yearLabel=new JLabel("年");
        yearLabel.setForeground(controlTextColor);
        result.add(yearLabel);
        monthSpin=new JSpinner(new SpinnerNumberModel(currentMonth,1,
                12,1));
        monthSpin.setPreferredSize(new Dimension(35,20));
        monthSpin.setName("Month");
        monthSpin.addChangeListener(this);
        result.add(monthSpin);
        JLabel monthLabel=new JLabel("月");
        monthLabel.setForeground(controlTextColor);
        result.add(monthLabel);
```

```java
        daySpin=new JSpinner(new SpinnerNumberModel(currentMonth,1,31,1));
        daySpin.setPreferredSize(new Dimension(35,20));
        daySpin.setName("Day");
        daySpin.addChangeListener(this);
        daySpin.setEnabled(false);
        daySpin.setToolTipText("请从下面的日历面板中进行选择哪一天!");
        result.add(daySpin);
        JLabel dayLabel=new JLabel("日");
        dayLabel.setForeground(controlTextColor);
        result.add(dayLabel);
        return result;
    }
    private JPanel createWeekAndDayPanal() {
        String colname[]={"日","一","二","三","四","五","六"};
        JPanel result=new JPanel();
        result.setFont(new Font("宋体",Font.PLAIN,12));
        result.setLayout(new GridLayout(7,7));
        result.setBackground(Color.white);
        JLabel cell;
        for (int i=0; i<7; i++) {
            cell=new JLabel(colname[i]);
            cell.setHorizontalAlignment(JLabel.RIGHT);
            if (i==0 || i==6) {
                cell.setForeground(weekendFontColor);
            } else {
                cell.setForeground(weekFontColor);
            }
            result.add(cell);
        }
        int actionCommandId=0;
        for (int i=0; i<6; i++) {
            for (int j=0; j<7; j++) {
                JButton numberButton=new JButton();
                numberButton.setBorder(null);
                numberButton.setHorizontalAlignment(SwingConstants.RIGHT);
                numberButton.setActionCommand(String.valueOf(actionCommandId));
                numberButton.addActionListener(this);
                numberButton.setBackground(palletTableColor);
                numberButton.setForeground(dateFontColor);
                if (j==0 || j==6) {
                    numberButton.setForeground(weekendFontColor);
                } else {
                    numberButton.setForeground(dateFontColor);
                }
                daysButton[i][j]=numberButton;
                result.add(numberButton);
```

```java
                actionCommandId++;
            }
        }
        return result;
    }
    public String getTextOfDateChooserButton() {
        return getText();
    }
        public void restoreTheOriginalDate() {
        String originalText=getOriginalText();
        setText(originalText);
    }
    private JPanel createButtonBarPanel() {
        JPanel panel=new JPanel();
        panel.setLayout(new java.awt.GridLayout(1,2));
        JButton ok=new JButton("确定");
        ok.addActionListener(new ActionListener() {
            public void actionPerformed(ActionEvent e) {
                //记忆原始日期时间
                initOriginalText(getTextOfDateChooserButton());
                //隐藏日历对话框
                dialog.setVisible(false);
            }
        });
        panel.add(ok);
        JButton cancel=new JButton("取消");
        cancel.addActionListener(new ActionListener() {

            public void actionPerformed(ActionEvent e) {
                //恢复原始日期时间
                restoreTheOriginalDate();
                //隐藏日历对话框
                dialog.setVisible(false);
            }
        });
        panel.add(cancel);
        return panel;
    }
    private JDialog createDialog(Frame owner) {
        JDialog result=new JDialog(owner,"日期时间选择",true);
        result.setDefaultCloseOperation(JDialog.HIDE_ON_CLOSE);
        result.getContentPane().add(this,BorderLayout.CENTER);
        result.pack();
        result.setSize(width,height);
        return result;
    }
    void showDateChooser(Point position) {
```

```java
            Frame owner=(Frame) SwingUtilities
                    .getWindowAncestor(DateChooserJButton.this);
            if (dialog==null || dialog.getOwner()!=owner) {
                dialog=createDialog(owner);
            }
            dialog.setLocation(getAppropriateLocation(owner,position));
            flushWeekAndDay();
            dialog.setVisible(true);
        }
        Point getAppropriateLocation(Frame owner,Point position) {
            Point result=new Point(position);
            Point p=owner.getLocation();
            int offsetX=(position.x+width)-(p.x+owner.getWidth());
            int offsetY=(position.y+height)-(p.y+owner.getHeight());
            if (offsetX>0) {
                result.x-=offsetX;
            }
            if (offsetY>0) {
                result.y-=offsetY;
            }
            return result;
        }
        private Calendar getCalendar() {
            Calendar result=Calendar.getInstance();
            result.setTime(getDate());
            return result;
        }
        private int getSelectedYear() {
            return ((Integer) yearSpin.getValue()).intValue();
        }
        private int getSelectedMonth() {
            return ((Integer) monthSpin.getValue()).intValue();
        }
        private int getSelectedHour() {
            return ((Integer) hourSpin.getValue()).intValue();
        }
        private int getSelectedMinite() {
            return ((Integer) minuteSpin.getValue()).intValue();
        }
        private int getSelectedSecond() {
            return ((Integer) secondSpin.getValue()).intValue();
        }
        private void dayColorUpdate(boolean isOldDay) {
            Calendar c=getCalendar();
            int day=c.get(Calendar.DAY_OF_MONTH);
            c.set(Calendar.DAY_OF_MONTH,1);
            int actionCommandId=day-2+c.get(Calendar.DAY_OF_WEEK);
```

```java
            int i=actionCommandId/7;
            int j=actionCommandId%7;
            if (isOldDay) {
                daysButton[i][j].setForeground(dateFontColor);
            } else {
                daysButton[i][j].setForeground(todayBackColor);
            }
        }
    }
    private void flushWeekAndDay() {
        Calendar c=getCalendar();
        c.set(Calendar.DAY_OF_MONTH,1);
        int maxDayNo=c.getActualMaximum(Calendar.DAY_OF_MONTH);
        int dayNo=2-c.get(Calendar.DAY_OF_WEEK);
        for (int i=0; i<6; i++) {
            for (int j=0; j<7; j++) {
                String s="";
                if (dayNo>=1 && dayNo<=maxDayNo) {
                    s=String.valueOf(dayNo);
                }
                daysButton[i][j].setText(s);
                dayNo++;
            }
        }
        dayColorUpdate(false);
    }
    //选择日期时的响应事件
    public void stateChanged(ChangeEvent e) {
        JSpinner source=(JSpinner) e.getSource();
        Calendar c=getCalendar();
        if (source.getName().equals("Hour")) {
            c.set(Calendar.HOUR_OF_DAY,getSelectedHour());
            setDate(c.getTime());
            return;
        }
        if (source.getName().equals("Minute")) {
            c.set(Calendar.MINUTE,getSelectedMinite());
            setDate(c.getTime());
            return;
        }
        if (source.getName().equals("Second")) {
            c.set(Calendar.SECOND,getSelectedSecond());
            setDate(c.getTime());
            return;
        }
```

```java
        dayColorUpdate(true);
        if (source.getName().equals("Year")) {
            c.set(Calendar.YEAR,getSelectedYear());
        } else if (source.getName().equals("Month")) {
            c.set(Calendar.MONTH,getSelectedMonth()-1);
        }
        setDate(c.getTime());
        flushWeekAndDay();
    }
    /*选择日期时的响应事件*/
    public void actionPerformed(ActionEvent e) {
        JButton source= (JButton) e.getSource();
        if (source.getText().length()==0) {
            return;
        }
        dayColorUpdate(true);
        source.setForeground(todayBackColor);
        int newDay=Integer.parseInt(source.getText());
        Calendar c=getCalendar();
        c.set(Calendar.DAY_OF_MONTH,newDay);
        setDate(c.getTime());
        daySpin.setValue(Integer.valueOf(newDay));
    }
}
```

17.2.7 罚款管理模块设计

1. 功能描述与分析

在图书管理系统中对于图书超期未还、图书损坏、图书丢失等现象要对读者进行罚款处理，在实际处理中要根据不同类型读者、不同罚款原因进行不同程度的罚款处理，同时对相应的图书进行登记处理，标注该书的状态。

2. 建立数据库表

该模块涉及的数据库表有罚款类型表 FineType、罚款信息表 FineInfo、借阅信息表 borrowInfo 和图书信息表 books，其中罚款类型表、罚款信息表结构如表 17-12 和表 17-13 所示。

表 17-12 罚款类型表结构

字段名	数据类型	长度	允许空	是否主键	含义
fine_id	int	4B	不允许	是	罚款类型编号
fine_kind	varchar	20 字符	允许		罚款类型
fine_reason	varchar	20 字符	允许		罚款原因

续表

字段名	数据类型	长度	允许空	是否主键	含义
fine_base	float	8B	允许		罚款基数
fine_times	float	8B	允许		罚款倍数

表 17-13 罚款信息表结构

字段名	数据类型	长度	允许空	是否主键	含义
fineinfo_id	int	4B	不允许	是	罚款信息编号
reader_id	int	4B	允许		读者编号
book_id	int	4B	允许		图书编号
fine_id	int	4B	允许		罚款类型
penalty	float	8B	允许		罚款金额
fine_date	datetime	8B	允许		罚款日期

3. 界面设计分析

图书管理员输入读者编号显示用户信息,输入图书编号显示图书相关信息,选择罚款类型和处理方式,对读者进行罚款处理;可以进行按时间段显示罚款信息汇总,查看罚款情况,如图 17-68 和图 17-69 所示。

图 17-68 "图书罚款处理"界面

图 17-69 "图书罚款汇总统计"界面

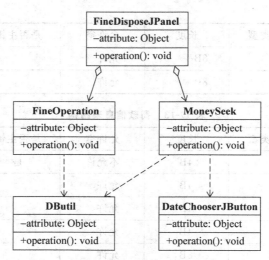

图 17-70 罚款管理相关类的 UML 图

4. 分析设计相关类

罚款管理操作中相关类的具体描述如表 17-14 所示。相关类如图 17-71～图 17-73 所示。

表 17-14 罚款管理相关类列表

类 名	UML 图
罚款处理操作类	**FineOperation** +addFine(): void +selectFine(): void +isFine(): void 图 17-71 "罚款处理操作类"类图
罚款处理界面类	**FineDisposeJPanel** −tfrid, tfrname, tfbid, tfbname: JTextField −tfbprice, tfbase, tftimes: JTextField −cbtype, cbreason: JComboBox −submit, cancel, query, st: JButton −table: JTable −fineop: FineOperation +FineDisposeJPanel() +actionPerformed(): void 图 17-72 "罚款处理界面类"类图
罚款统计界面类	**MoneySeek** −start, stop: DateChooserJButton −seek, sum, dayin: JButton −jtable: JTable +MoneySeek() +actionPerformed(): void 图 17-73 "罚款统计界面类"类图

5. 相关类主要代码

该模块设计涉及类的主要代码如下。

罚款处理界面类 FineDisposeJPanel：

```
public class FineDisposeJPanel extends JFrame {
    JPanel top,center,bottom,bp,psum,p;
    JLabel rid,rname,bid,bname,bprice,finebase,finetimes,type,reason;
    JTextField tfrid,tfrname,tfbid,tfbname,tfbprice,tfbase,tftimes;
    JComboBox cbtype,cbreason;
    JButton submit,cancel,query,st;
    JTable table;
    Object a[][]=new Object[8][7];
    Object b[]={"读者编号","姓名","图书编号","图书名称","罚款类型","还书时间","罚款金额"};
    JScrollPane jsp;
    FineDisposeJPanel() {
        setTitle("罚款处理");
        rid=new JLabel("读者编号：");
        rname=new JLabel("读者姓名：");
        tfrid=new JTextField(8);
        tfrname=new JTextField(8);
        tfrname.setEditable(false);
        top=new JPanel();
        center=new JPanel();
        top.add(rid);
        top.add(tfrid);
        top.add(rname);
        top.add(tfrname);
        bid=new JLabel("图书编号：");
        bname=new JLabel("图书名称：");
        bprice=new JLabel("价格：");
        tfbid=new JTextField(8);
        tfbname=new JTextField(8);
        tfbprice=new JTextField(6);
        center.add(top);
        center.add(bid);
        center.add(tfbid);
        center.add(bname);
        center.add(tfbname);
        center.add(bprice);
        center.add(tfbprice);
        tfbname.setEditable(false);
        tfbprice.setEditable(false);
        type=new JLabel("  罚款类型：");
        reason=new JLabel("  罚款原因：");
```

```
cbtype=new JComboBox();
cbtype.addItem("超期未还");
cbtype.addItem("图书损坏");
cbtype.addItem("图书丢失");
cbreason=new JComboBox();
cbreason.addItem("超期未还");
cbreason.addItem("损坏 10%~20%");
cbreason.addItem("损坏 30%~50%");
cbreason.addItem("损坏 60%~100%");
cbreason.addItem("图书丢失");
finebase=new JLabel("  罚款基数:");
tfbase=new JTextField(10);
finetimes=new JLabel("  罚款倍数:");
tftimes=new JTextField(10);
bottom=new JPanel();
bottom.add(type);
bottom.add(cbtype);
bottom.add(reason);
bottom.add(cbreason);
bottom.add(finebase);
bottom.add(tfbase);
bottom.add(finetimes);
bottom.add(tftimes);
p=new JPanel();
p.setLayout(new BorderLayout());
p.add(center,BorderLayout.NORTH);
p.add(bottom,BorderLayout.CENTER);
for(int i=0;i<8;i++)
{ for(int j=0;j<7;j++)
    { a[i][j]=""; }
}
table=new JTable(a,b);
jsp=new JScrollPane(table);
submit=new JButton("罚款处理");
cancel=new JButton("取消");
query=new JButton("信息汇总");
st=new JButton("罚款统计");
bp=new JPanel();
bp.add(submit);
bp.add(cancel);
bp.add(query);
bp.add(st);
add(p,BorderLayout.NORTH);
add(jsp,BorderLayout.CENTER);
add(bp,BorderLayout.SOUTH);
```

```java
        setSize(850,300);
        setVisible(true);
    }
}
```

罚款统计界面类 MoneySeek：

```java
public class MoneySeek extends JFrame implements ActionListener{
    DateChooserJButton start=new DateChooserJButton();
    DateChooserJButton stop=new DateChooserJButton();
    JTable jt;
    JButton seek,sum,dayin;
    JLabel jl;
    public MoneySeek() {
        super("罚款统计");
        start.setBorder(BorderFactory.createLineBorder(new Color(51,255,51)));
        stop.setBorder(BorderFactory.createLineBorder(new Color(51,255,51)));
        seek=new JButton("查询");
        seek.addActionListener(this);
        dayin=new JButton("打印");
        dayin.addActionListener(this);
        sum=new JButton("总罚款额");
        sum.addActionListener(this);
        jt=new JTable(new BorrowTableModel(1));
        JScrollPane js=new JScrollPane(jt);
        js.setPreferredSize(new Dimension(600,350));
        add(new JLabel("开始时间："));
        add(start);
        add(new JLabel("截止时间："));
        add(stop);
        add(seek);
        add(dayin);
        add(sum);
        add(js);
        jl=new JLabel();
        jl.setFont(new Font("宋体",Font.BOLD,30));
        add(jl);
        setLayout(new FlowLayout());
        setVisible(true);
        setBounds(300,100,700,500);
    }
    public void actionPerformed(ActionEvent e) {
        if(e.getSource()==seek){
            jt.setModel(new BorrowTableModel(start.getText(),stop.getText(),1));
        }
        if(e.getSource()==dayin){
```

```
            new BorrowManage().moneyReport();
        }
        if(e.getSource()==sum){
            jl.setText("一共罚款："+new BorrowManage().sumMoney()+"元。");
        }
    }
}
```

在归还图书时进行罚款处理，FineOperation 类主要代码如下：

```
Date d1=sdf.parse(startTime,pos1);
Date d2=sdf.parse(time,pos2);
    day=(int)((d1.getTime()-d2.getTime())/86400000);
    day=day>0? day:day*-1;
    sql="select * from reader where reader_id="+readerID;
    rs=sm.executeQuery(sql);
    if(rs.next()){
        type=rs.getString("type");
    }
    DBUtil.closeCon(con);
    penalty=new CheckUser().money(type,day);
      if(type.equals("学生")){
            if(day>30){
                penalty=(day-30) * new CheckUser().money(type);
            }
      }
      if(type.equals("教师")){
          if(day>60){
              penalty=(day-60) * new CheckUser().money(type);
          }
      }
    con=DBUtil.getConnection();
    sm=con.createStatement();
    if(penalty>0){
        int a=JOptionPane.showConfirmDialog(jop,"因你超过了还书期限,需要交纳"
+penalty+"元罚金。");
    if(a==0){
        con.setAutoCommit(false);
        sql="update borrowInfo set stopTime='"+time+"',penalty="+penalty+
            " where reader_id="+readerID+" and book_id="+bookID+"";
        row=sm.executeUpdate(sql);
        sql="update books set stock=stock+1 where book_id="+bookID;
        row=sm.executeUpdate(sql)+row;
        con.commit();
        con.setAutoCommit(true);
    }
```

17.2.8 报表打印模块设计

1. 功能描述与分析

存储在数据库中的数据,可以显示在界面上,也可以以报表的形式打印输出。使用 ireport 开源报表软件 JasperReport 的可视化设计工具,设计生成一个 Jasperreports 报表文件,把报表文件 DTD 定义的 xml 格式的源文件编译成一个 jasper 类型的文件,jasper 文件可以在应用程序中被加载生成最终的报表。

Jasperreports 报表文件被垂直分成若干个部分,每一个部分称为 band。具体划分如下。

(1) Title Band:只在整个报表的第一页的最上面部分显示。

(2) pageHeader Band:pageHeader 中的内容将会在整个报表中的每一个页面中都会出现,显示位置在页面的上部,如果是报表的第一页,pageHeader 中的内容将显示在 Title Band 下面。

(3) pageFooter Band:显示在每页页面的最下端。

(4) Detail Band:报表内容段,重复出现的内容,Detail 段中的内容每页都会出现。

(5) columnHeader Band:表头段,在报表内容的上方设计报表的列头。

(6) columnFooter Band:表尾段,在报表内容的下方显示的列尾。

(7) Summary Band:表格的合计段,出现在整个报表的最后一页中的 Detail Band 的后面,一般用来统计报表中某一个或某几个字段的合计值。

使用立方 ireport 软件操作过程:打开立方 ireport 软件,选择"数据"菜单下的"数据源连接",先建立数据连接,操作界面如图 17-74 所示。

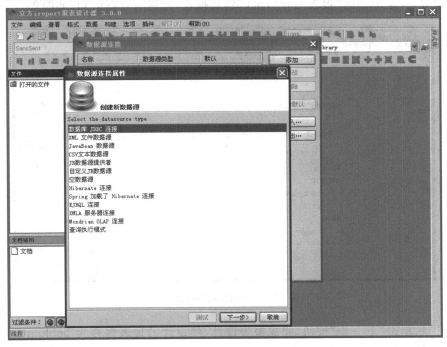

图 17-74 使用 ireport 设计器创建数据源界面

单击"下一步"按钮,创建数据源,填写数据源名称、连接数据库的驱动程序、连接数据库的地址和数据库名以及数据库连接的用户名和密码,完成后单击"测试"按钮,测试连接是否成功,如图 17-75 所示。

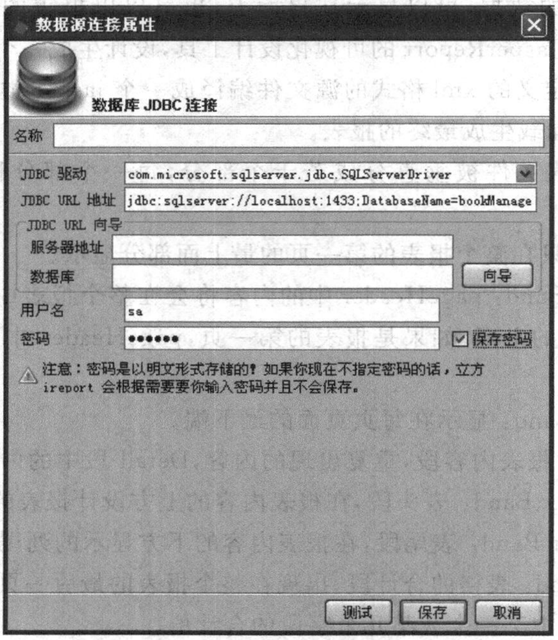

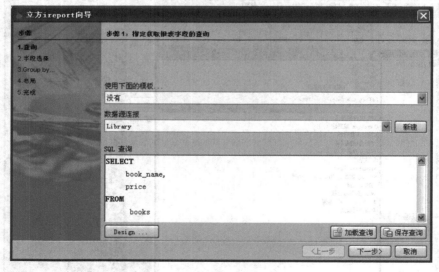

图 17-75 数据源连接设置界面

利用"文件"菜单中的"报表向导",可以创建报表,先编写查询语句,如图 17-76 所示。

图 17-76 创建报表向导界面

根据提示单击"下一步"按钮,完成报表设计,将设计的报表保存为 book.jrxml,看到设计界面,如图 17-77 所示。在设计界面中可以进行修改,完成报表如图 17-78 所示。

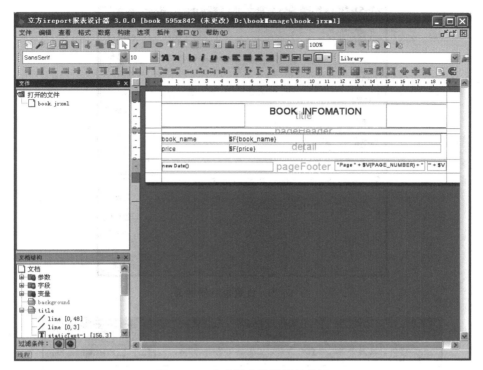

图 17-77 完成报表设计后的界面

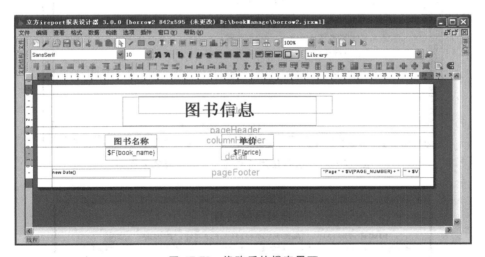

图 17-78 修改后的报表界面

在"构建"菜单先可以选择查看方式,选择不同查看方式,使用不同查看器,其设置方式是在"选项"菜单中的"设置"命令中,进行设置查看器,如图 17-79 所示。

选择"构建"菜单下"执行报表(空数据源)"或"执行报表(使用活动连接)",可以查看报表设计的执行效果,如图 17-80 所示。

2. 界面设计分析

可以设计打印相关信息,如设计打印学生查询图书的信息,操作界面如图 17-81 所示。

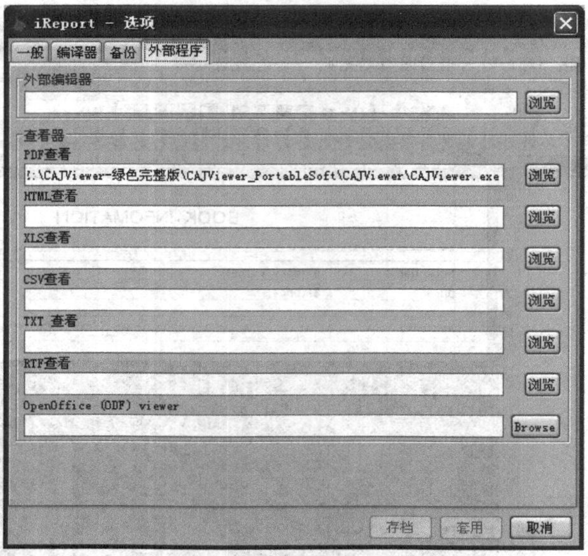

图 17-79 设置查看器界面

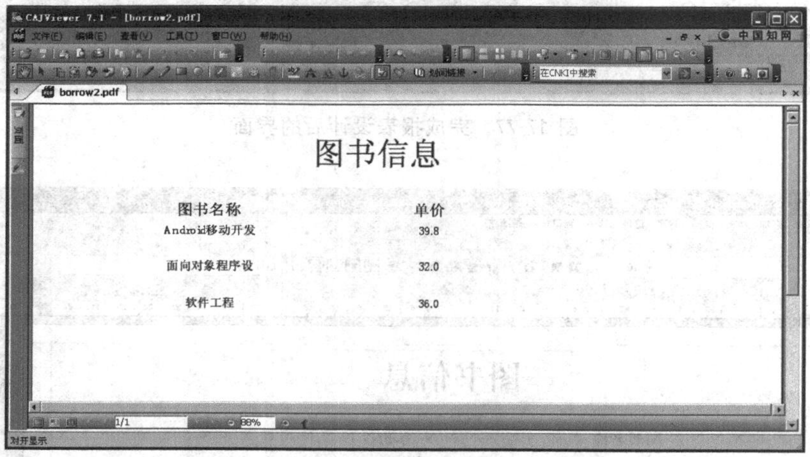

图 17-80 "执行使用活动连接"后报表效果界面

图 17-81 "查询借阅信息"中设置打印按钮

单击“打印”按钮，以 PDF 格式打印预览报表格式，如图 17-82 所示。

图书借阅信息

读者姓名	读者类型	图书名称	联系方式	借书时间	还书时间	罚金
梁健	学生	面向对象程序设	13525655261	2014-03-04 00:	2014-03-10 00:	1.2
孙旭	学生	Android移动开发	13525452747	2014-09-01 00:	2014-10-01 00:	0.0
孙旭	学生	软件工程	13525452747	2014-09-01 00:	2014-10-10 00:	0.0
张涵予	学生	面向对象程序设	18660045982	2014-09-11 00:	2014-10-05 00:	0.0
王爱娜	学生	纳税申报实物	13105426863	2014-12-11 00:	null	0.0
陈晓峰	教师	把盏话茶	13542048201	2014-12-11 00:	null	0.0
陈晓峰	教师	纳税风险控制手	13542048201	2014-12-11 00:	null	0.0

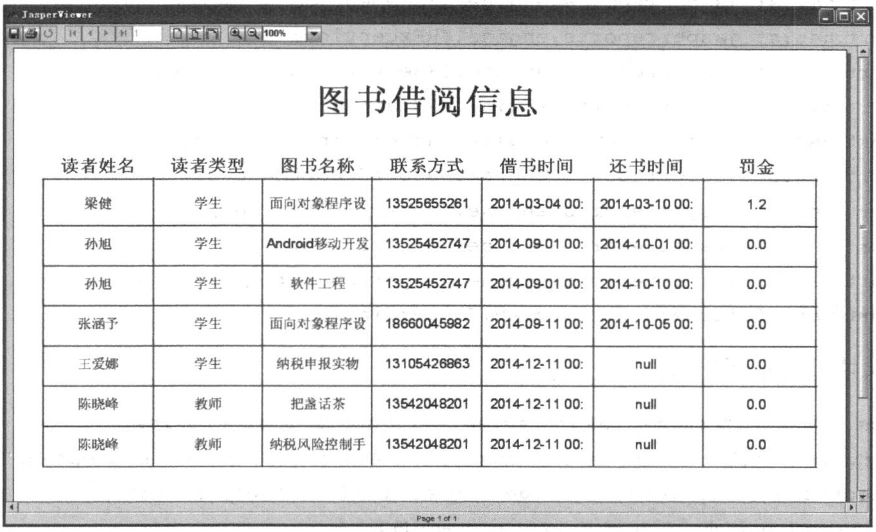

图 17-82　“图书借阅信息”打印效果预览

3. 分析设计相关类

打印设计实现代码。

1）导入相关打印包

在程序中通过导入打印报表相关包，在 Project 菜单下的 Properties 中，选择 Java Build Path，单击 Add External JARs 按钮，如图 17-83 所示，导入 commons-beanutils-1.8.0.jar、commons-collections-2.1.1.jar、commons-digester-2.1.jar、commons-javaflow-20060411.jar、commons-logging-1.1.1.jar、iText-2.1.7.js1.jar 和 jasperreports-5.0.0.jar。

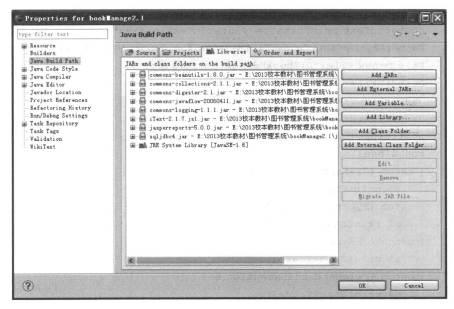

图 17-83　在 Eclipse 中导入相关包

```java
import net.sf.jasperreports.engine.JRBand;
import net.sf.jasperreports.engine.JRElement;
import net.sf.jasperreports.engine.JRException;
import net.sf.jasperreports.engine.JRResultSetDataSource;
import net.sf.jasperreports.engine.JRStaticText;
import net.sf.jasperreports.engine.JasperCompileManager;
import net.sf.jasperreports.engine.JasperExportManager;
import net.sf.jasperreports.engine.JasperFillManager;
import net.sf.jasperreports.engine.JasperPrint;
import net.sf.jasperreports.engine.JasperReport;
import net.sf.jasperreports.view.JasperViewer;
```

2）设计打印报表信息

```java
Class.forName("com.microsoft.sqlserver.jdbc.SQLServerDriver");
String url="jdbc:sqlserver://localhost:1433;databasename=bookmanage";
Connection con=DriverManager.getConnection(url,"sa","123456");
    Statement stmt=con.createStatement();
    String sql="select reader_name,reader.type,book_name,books.type,phone,
    startTime,stopTime,penalty
    from books,reader,borrowInfo
    where books.book_id=borrowInfo.book_id and reader.reader_id=
    borrowInfo.reader_id";
    ResultSet rs=stmt.executeQuery(sql);
    JRResultSetDataSource ds=new JRResultSetDataSource(rs);
    JasperReport report=JasperCompileManager
        .compileReport("D:/bookManage/borrow1.jrxml");
    JRBand header=report.getPageHeader();
    JRElement element=header.getElementByKey("printDate");
    JRStaticText printDateSt=(JRStaticText)element;
    JasperPrint print=JasperFillManager.fillReport(report,new HashMap(),ds);
    File f=new File("D:/bookManage/borrow1.pdf");
    OutputStream output=new FileOutputStream(f);
    JasperExportManager.exportReportToPdfStream(print,output);
    JasperViewer.viewReport(print);
    stmt.close();
    con.close();
} catch (ClassNotFoundException e1) {
    e1.printStackTrace();
} catch (SQLException e2) {
    e2.printStackTrace();
} catch (JRException e3) {
    e3.printStackTrace();
} catch (FileNotFoundException e4) {
    e4.printStackTrace();
}
```

17.2.9 帮助管理模块设计

1. 功能描述与分析

在帮助菜单下,设计图书馆管理系统的关于、帮助、图书馆规则提示等界面。关于面板提供关于系统的信息,可以直接设计一个面板界面实现,可以在面板上设置背景图片,使界面美化。

图书馆规则帮助用户了解读者借阅规则和图书馆管理办法,该功能利用文件输入实现,将图书馆规则编写在文档中,通过数据输入流将相关规则读入到规则面板中,显示出来,为了提高数据读取速度,使用缓冲输入流,读入的数据放在滚动面板中。

利用 Easy CHM 制作图书管理系统帮助手册,当用户有问题时,可以查阅帮助手册,了解项目功能概况,也可以根据不同目录标题找到问题解决方案,让新用户快速上手,方便操作系统。

2. 界面设计分析

(1) "关于"界面设计如图 17-84 所示。

(2) "图书馆规则"界面设计如图 17-85 所示。

图 17-84 "关于"界面设计

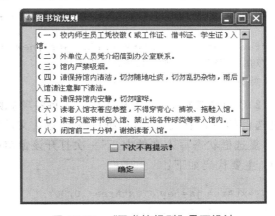

图 17-85 "图书馆规则"界面设计

(3) 帮助手册界面设计如图 17-86 所示。

3. 分析设计相关类

(1) 在"关于"界面中利用图片作为背景,然后添加其他控件,实现界面设计,主要代码如下:

```
background=new ImageIcon("ico/background.jpg");    //背景图片
JLabel label=new JLabel(background);               //把背景图片显示在一个标签里面
//把标签的大小位置设置为图片刚好填充整个面板
label.setBounds(0,0,background.getIconWidth(),background.getIconHeight());
//把内容窗格转化为 JPanel,否则不能用方法 setOpaque()来使内容窗格透明
imagePanel=(JPanel) frame1.getContentPane();
imagePanel.setOpaque(false);
//内容窗格默认的布局管理器为 BorderLayout
imagePanel.setLayout(new FlowLayout());
frame1.getLayeredPane().setLayout(null);
//把背景图片添加到分层窗格的最底层作为背景
```

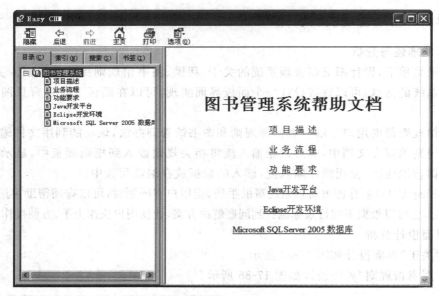

图 17-86 帮助手册界面

```
frame1.getLayeredPane().add(label,new Integer(Integer.MIN_VALUE));
//在内容窗格中添加其他各种控件
imagePanel.add(各类控件);
```

(2)"图书馆规则"面板利用输入流将各项规则读入到滚动面板中;该面板是否打开,根据复选框的设置有关,而复选框的状态是从文件读取来的,每次用户单击复选框进行关闭后,将复选框状态写入文件中,下一次打开该窗口时,先从文件中读取状态,决定是否显示该窗口,主要代码如下:

```
JTextArea ja=new JTextArea(10,30);
FileReader fr=null;
BufferedReader br=null;
ja.setLineWrap(true);
ja.setEditable(false);
js=new JScrollPane(ja);
try {
    fr=new FileReader("D:/bookManage/rule.txt");
    br=new BufferedReader(fr);
    while((s=br.readLine())!=null){
        ja.append(s);
        ja.append("\n");
    }
    br.close();
    fr.close();
} catch (FileNotFoundException e) {
    e.printStackTrace();
} catch (IOException e) {
    e.printStackTrace();
```

```
    }
```

(3) 帮助文档制作与调用。

利用 Easy CHM 制作帮助文档操作步骤如下。

① 先将一些相关问题及解决方案制作成网页,保存格式为 html 或 htm,保存在同一文件夹中。

② 打开 Easy CHM,单击"文件"菜单下的"新建"图标,新建工程目录,如图 17-87 和图 17-88 所示。

图 17-87　打开 Easy CHM 软件主界面

图 17-88　"新工程目录"对话框

③ 单击"浏览"按钮将文件夹中的文件导入,根据目录结构和文档顺序,调整各级目录次序和目录名称。单击"文件"菜单中的"编译指定工程"选项,将该工程进行编写,生成 help.chm 文件,如图 17-89 所示。

④ 在菜单中要调用该 chm 文件,使用语句

```
Runtime.getRuntime().exec("cmd /c start help.chm");
```

可以调用 help.chm 文件,在程序中显示帮助手册。

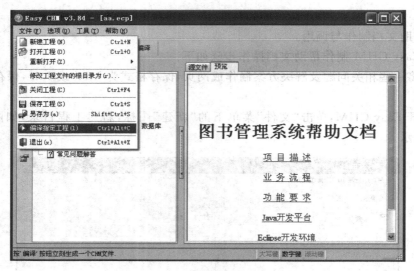

图 17-89 编译工程获得 CHM 文件

17.2.10 主界面管理模块设计

1. 功能描述与分析

设计完成各个模块的开发与实现,设计图书管理系统的主界面,在主界面中设计相应菜单项和命令按钮,完成功能模块的调用。

2. 界面设计分析

(1) 进入图书管理系统主界面后,如图 17-90 所示。

图 17-90 图书管理系统主界面

(2) 对各个菜单进行设计，如图 17-91～图 17-97 所示。

图 17-91　"文件"菜单设计

图 17-92　"读者"菜单设计

图 17-93　"图书"菜单设计

图 17-94　"借阅"菜单设计

图 17-95　"信息统计"菜单设计

图 17-96　"系统维护"菜单设计

(3) 在主界面上面设计一个 top 面板，放置图片，显示图片信息；设计创建 JToolBar 类对象，放置各个按钮，对齐排列，并显示登录用户账号、登录时间；下面设计 bottom 面板，当单击不同按钮时，更新面板中显示内容，显示效果如图 17-98～图 17-104 所示。

图 17-97 "帮助"菜单设计

图 17-98 "首页"显示信息界面

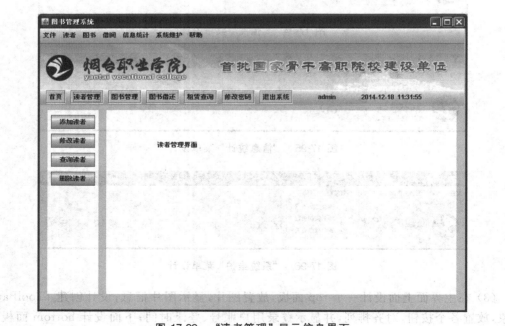

图 17-99 "读者管理"显示信息界面

图 17-100 "图书管理"显示信息界面

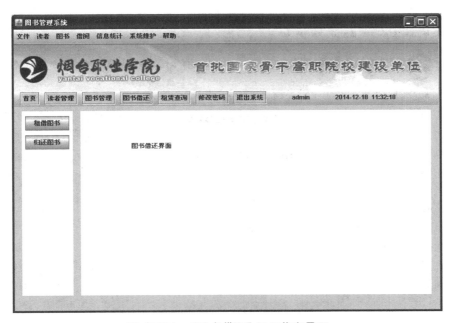

图 17-101 "图书借还"显示信息界面

图 17-102 "租赁查询"显示信息界面

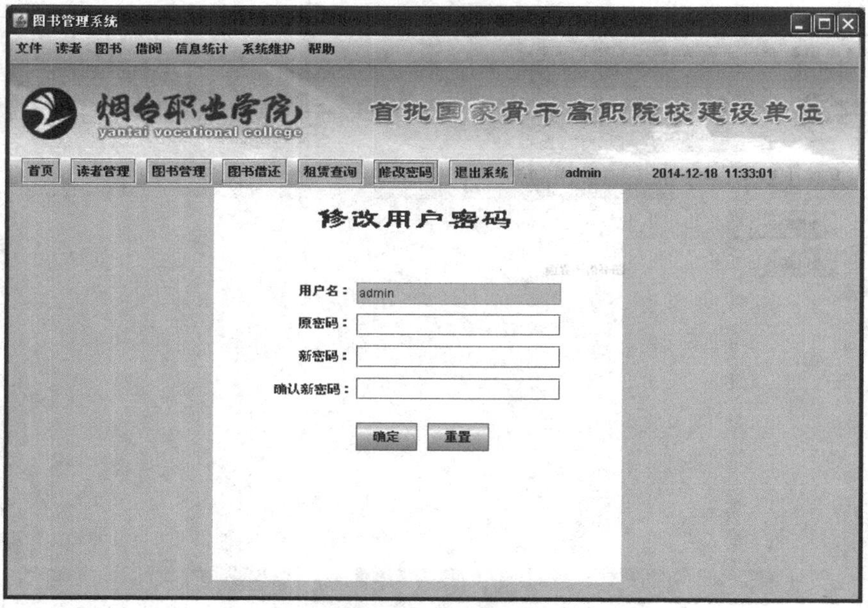

图 17-103 "修改密码"显示信息界面

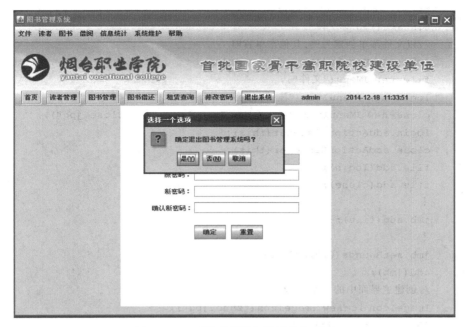

图 17-104 "退出系统"显示信息界面

3．分析设计相关类

在主界面类中调用前面各个类，其相互关系如图 17-105 所描述。

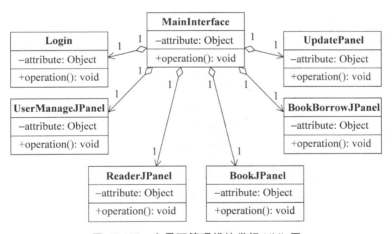

图 17-105 主界面管理模块类间 UML 图

4．程序代码

该程序主界面类主要代码如下：

```
public class MainInterface extends JFrame implements ActionListener,
WindowListener,MouseListener {
…//声明窗体中的各类对象
    public MainInterface() {
    super("图书管理系统");
```

```
t=new Timer(1000,this);
t.start();
//以文件菜单创建为例,其他菜单省略
jmb=new JMenuBar();
file=new JMenu("文件");
login=new JMenuItem("注销",new ImageIcon("ico/46.png"));
close=new JMenuItem("退出",new ImageIcon("ico/close.jpg"));
login.addActionListener(this);
close.addActionListener(this);
file.add(login);
file.add(close);
    ⋮
jmb.add(file);
    ⋮
jmb.setBounds(0,0,830,30);
add(jmb);
//创建主界面中的top面板
ImageIcon ic=new ImageIcon("zhuo.jpg");
JLabel jl=new JLabel(ic);
jpTop=new JPanel();
jpTop.add(jl);
jpTop.setBounds(0,25,830,100);
add(jpTop);
//创建工具条
JToolBar jtb=new JToolBar();
jbMain=new JButton("首页");
jbMain.addActionListener(this);
jbUser=new JButton("读者管理");
jbUser.addActionListener(this);
jbBook=new JButton("图书管理");
jbBook.addActionListener(this);
jbBorrow=new JButton("图书借还");
jbBorrow.addActionListener(this);
jbSeek=new JButton("租赁查询");
jbSeek.addActionListener(this);
jbUpdate=new JButton("修改密码");
jbUpdate.addActionListener(this);
jbClose=new JButton("退出系统");
jbClose.addActionListener(this);
SimpleDateFormat sdf=new SimpleDateFormat("yyyy-MM-dd HH:mm:ss");
Date d=new Date();
String time=sdf.format(d);
jll=new JLabel(time);
jtb.addSeparator();
jtb.add(jbMain);
```

```java
jtb.addSeparator();
jtb.add(jbUser);
jtb.addSeparator();
jtb.add(jbBook);
jtb.addSeparator();
jtb.add(jbBorrow);
jtb.addSeparator();
jtb.add(jbSeek);
jtb.addSeparator();
jtb.add(jbUpdate);
jtb.addSeparator();
jtb.add(jbClose);
jtb.addSeparator();
jluser=new JLabel(status);
jluser.setForeground(Color.blue);
jluser.addMouseListener(this);
jtb.add(jluser);
jtb.addSeparator();
jtb.addSeparator();
jtb.addSeparator();
jtb.add(jl1);
jtb.setFloatable(false);
jtb.setBounds(0,125,830,30);
add(jtb);
//设置下面面板
all=new JPanel();
all.setLayout(null);
ic=new ImageIcon("paih.jpg");
JLabel jl1=new JLabel(ic);
jpTop1=new JPanel();
jpTop1.add(jl1);
jpTop1.setBounds(0,0,830,55);
jt=new JTable();
init();
jt.setDefaultRenderer(Object.class,new EvenOddRenderer());
jpDown=new JPanel();
JScrollPane js=new JScrollPane();
js.setViewportView(jt);
Dimension dm=new Dimension(800,300);
js.setPreferredSize(dm);
jpDown.add(js);
jpDown.setBounds(10,55,800,310);
ic=new ImageIcon("1.gif");
JLabel jl2=new JLabel(ic);
jpDown1=new JPanel();
```

```java
        jpDown1.add(jl2);
        jpDown1.setBounds(0,365,830,30);
        all.add(jpTop1);
        all.add(jpDown);
        all.add(jpDown1);
        all.setBounds(0,150,830,400);
        add(all);
        jp.setLayout(null);
        jlPrompt.setBounds(100,20,100,100);
        jp.add(jlPrompt);
        jp.setBounds(130,15,685,370);
        jp.setBackground(Color.white);
        readers.setBounds(0,150,830,400);
        add(readers);
        books.setBounds(0,150,830,400);
        add(books);
        bookBorrow.setBounds(0,150,830,400);
        add(bookBorrow);
        seekBorrow.setBounds(60,150,700,400);
        add(seekBorrow);
        up.setBounds(200,150,400,400);
        add(up);
        addWindowListener(this);
        setResizable(false);
        setLayout(null);
        setBounds(250,50,840,600);
        setVisible(true);
        setDefaultCloseOperation(JFrame.DO_NOTHING_ON_CLOSE);
    }
    //事件监听处理
    public void actionPerformed(ActionEvent e) {
        if(e.getSource()==jbMain||e.getSource()==cLend){
            windowInit();
            init();
            all.setVisible(true);
        }
        if(e.getSource()==jbUser){
            windowInit();
            readers.add(jp);
            jlPrompt.setText("读者管理界面");
            jp.setVisible(true);
            readers.setVisible(true);
        }
        if(e.getSource()==jbBook){
            windowInit();
```

```java
        books.add(jp);
        jlPrompt.setText("图书管理界面");
        jp.setVisible(true);
        books.setVisible(true);
    }
    if(e.getSource()==jbBorrow){
        windowInit();
        bookBorrow.add(jp);
        jlPrompt.setText("图书借还界面");
        jp.setVisible(true);
        bookBorrow.setVisible(true);
    }
    if(e.getSource()==jbSeek||e.getSource()==borrowSeek){
        windowInit();
        seekBorrow.init();
        seekBorrow.setVisible(true);
    }
    if(e.getSource()==jbUpdate){
        windowInit();
        up.setVisible(true);
    }
    if(e.getSource()==rAdd){
        windowInit();
        readers.setVisible(true);
        ReaderJPanel.arj.setVisible(true);
        ReaderJPanel.arj.init();
    }
    if(e.getSource()==rUpdate){
        windowInit();
        readers.setVisible(true);
        ReaderJPanel.urj.setVisible(true);
        ReaderJPanel.urj.init();
    }
    if(e.getSource()==rSeek){
        windowInit();
        readers.setVisible(true);
        ReaderJPanel.srj.setVisible(true);
        ReaderJPanel.srj.init();
    }
    if(e.getSource()==rDelete){
        windowInit();
        readers.setVisible(true);
        ReaderJPanel.drj.setVisible(true);
        ReaderJPanel.drj.init();
    }
```

```java
if(e.getSource()==bAdd){
    windowInit();
    books.setVisible(true);
    BookJPanel.abj.setVisible(true);
    BookJPanel.abj.init();
}
if(e.getSource()==bUpdate){
    windowInit();
    books.setVisible(true);
    BookJPanel.ubj.setVisible(true);
    BookJPanel.ubj.init();
}
if(e.getSource()==bDelete){
    windowInit();
    books.setVisible(true);
    BookJPanel.dbj.setVisible(true);
    BookJPanel.dbj.init();
}
if(e.getSource()==bSeek){
    windowInit();
    books.setVisible(true);
    BookJPanel.sbj.setVisible(true);
    BookJPanel.sbj.init();
}
if(e.getSource()==bLend){
    windowInit();
    bookBorrow.setVisible(true);
    BookBorrowJPanel.lbj.setVisible(true);
    BookBorrowJPanel.lbj.init();
}
if(e.getSource()==bReturn){
    windowInit();
    bookBorrow.setVisible(true);
    BookBorrowJPanel.rbj.setVisible(true);
    BookBorrowJPanel.rbj.init();
}
if(e.getSource()==cBook){
    new BookCount();
}
if(e.getSource()==sRType){
    if(CheckUser.user_type.equals("管理员")){
        new TypeManageJFrame();
    }else{
        JOptionPane.showMessageDialog(jop,"权限不足!");
    }
```

```java
        }
        if(e.getSource()==sBType){
            if(CheckUser.user_type.equals("管理员")){
                new BookTypeManage();
            }else{
                JOptionPane.showMessageDialog(jop,"权限不足!");
            }
        }
        if(e.getSource()==sMoney){
            if(CheckUser.user_type.equals("管理员")){
                new MoneyManage();
            }else{
                JOptionPane.showMessageDialog(jop,"权限不足!");
            }
        }
        if(e.getSource()==help1){
            try {
                Runtime.getRuntime().exec("cmd /c start bookManageHelp.chm");
            } catch (IOException e1) {
                //TODO Auto-generated catch block
                e1.printStackTrace();
            }
        }
        if(e.getSource()==guanyu){
            new GuanYu();
        }
        if(e.getSource()==guize){
            new Rule();
        }
        if(e.getSource()==login){
            status=null;
            this.setVisible(false);
            new Login();
        }
        if(e.getSource()==jbClose||e.getSource()==close){
            int i=JOptionPane.showConfirmDialog(jop,"确定退出图书管理系统吗?");
            if(i==0){
                t.stop();
                status=null;
                System.exit(0);
            }
        }
        if(e.getSource()==t){
            timer();
        }
```

```java
        }
        public void timer(){
            SimpleDateFormat sdf=new SimpleDateFormat("yyyy-MM-dd HH:mm:ss");
            Date d=new Date();
            jll.setText(sdf.format(d));
        }
        public void init(){
            jt.setModel(new BookTableModel());
        }
        public void windowActivated(WindowEvent e) {}
        public void windowClosed(WindowEvent e) {}
        public void windowClosing(WindowEvent e) {
            int i=JOptionPane.showConfirmDialog(jop,"确定退出图书管理系统吗?");
            if(i==0){
                t.stop();
                status=null;
                System.exit(0);
            }
        }
        public void windowDeactivated(WindowEvent e) {}
        public void windowDeiconified(WindowEvent e) {}
        public void windowIconified(WindowEvent e) {}
        public void windowOpened(WindowEvent e) {
            File f=new File("D:/bookManage/guize.txt");
            if(f.exists()){
                try {
                    FileReader in=new FileReader(f);
                    BufferedReader reader=new BufferedReader(in);
                    try {
                        if(reader.readLine().equals("false")){
                            new Rule();
                        }else{
                        }
                        in.close();
                        reader.close();
                    } catch (IOException e1) {
                        //TODO Auto-generated catch block
                        e1.printStackTrace();
                    }
                } catch (FileNotFoundException e1) {
                    //TODO Auto-generated catch block
                    e1.printStackTrace();
                }
            }else{
                new Rule();
            }
        }
```

```java
    public void mouseClicked(MouseEvent e) {
        all.setVisible(false);
        readers.setVisible(false);
        books.setVisible(false);
        bookBorrow.setVisible(false);
        seekBorrow.setVisible(false);
        up.setVisible(true);
    }
    public void mouseEntered(MouseEvent e) {
        jluser.setForeground(Color.darkGray);
        jluser.setCursor(Cursor.getPredefinedCursor(Cursor.HAND_CURSOR));
    }
    public void mouseExited(MouseEvent e) {
        jluser.setForeground(Color.blue);
    }
    public void mousePressed(MouseEvent e) {}
    public void mouseReleased(MouseEvent e) {}
    public void windowInit(){
        seekBorrow.setVisible(false);
        up.setVisible(false);
        all.setVisible(false);
        books.setVisible(false);
        bookBorrow.setVisible(false);
        readers.setVisible(false);
        ReaderJPanel.arj.setVisible(false);
        ReaderJPanel.urj.setVisible(false);
        ReaderJPanel.drj.setVisible(false);
        ReaderJPanel.srj.setVisible(false);
        BookJPanel.abj.setVisible(false);
        BookJPanel.ubj.setVisible(false);
        BookJPanel.sbj.setVisible(false);
        BookJPanel.dbj.setVisible(false);
        BookBorrowJPanel.lbj.setVisible(false);
        BookBorrowJPanel.rbj.setVisible(false);
    }
}
```

17.3 系统发布与总结

17.3.1 项目打包

在本项目中,最终需要打入 jar 包的文件有很多,其中包括如下内容。
(1) 运行应用程序所必需的 class 文件。
(2) SQL Server 2005 JDBC 驱动程序 sqljdbc.jar。
(3) 生成 PDF 文件时所需要的 iText 远东字体包文件 iTextAsian.jar。
(4) 源报表工具 JasperReports 2.0.5 工具包。

使用 Eclipse 中的打包功能，将整个项目打包，发布成.jar 文件，让用户直接使用该程序。在该项目 bookManage2.1 中右击，选择 export 命令，在其中选择 Runnable JAR file 命令，如图 17-106 所示。

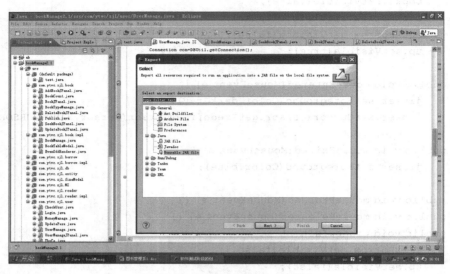

图 17-106　项目打包 export 界面

单击 Next 按钮，在 Launch configuration 下的文本框中设置项目启动的主程序，Export destination 下的文本框中设置打包后文件位置，然后单击 Finish 按钮，即可完成项目的打包，如图 17-107 所示。

图 17-107　打包项配置界面

17.3.2 项目总结

1. 需求分析小结

需求分析是软件开发的开始阶段,在软件开发中占有非常重要的地位。需求分析就是分析软件用户的需求是什么,如果需求分析出了差错,致使投入了大量的人力、物力、财力和时间开发出来的软件最后却不能满足用户的需求,那会造成巨大的损失。

在项目需求中说明了图书管理系统的重要性和必要性。在需求描述中,以学校图书馆为例,描述了从读者注册到读者借阅的整个过程。通过对需求分析的了解,让学生了解图书馆业务知识,培养学生对系统的开发兴趣,便于以后的系统开发进展顺利。

在系统分析设计中,通过需求描述抽取系统功能,根据功能进行模块划分以下几类:图书管理、读者管理、图书借阅、罚款管理和报表输出。

2. 概要设计小结

概要设计主要是将用户需求转化为未来的系统,概要设计中需要明确下面3点。

(1) 技术架构:概要设计中需要明确系统采用的技术是什么。例如,用什么开发技术、框架、数据库等。

(2) 功能模块划分:要进行进一步开发,功能模块的细化必须在概要设计中完成。概要设计中功能模块需要说明模块中用到了哪些类,这些类的功能是什么,类与类之间的调用关系。模块的输入数据是什么,输出数据是什么同时也要在概要设计的模块划分中进行明确。

(3) 数据库设计:在概要设计阶段设计出数据库有哪些表,表的具体结构、表与表之间的关系都要在概要设计阶段完成。

概要设计确定了系统的大方向以及系统的技术构成。详细设计是以概要设计为基础的一个细化过程。

会员管理系统的概要设计包括了以上所提到的方面。

(1) 技术架构:会员管理系统采用 C/S(Client/Server)结构,系统的程序架构将会分为三层,配合 SQL Server 2005 数据库。在设计中采用 MVC 的架构思想,其中 View 包中包含系统所有界面窗体,属于表示层。Model 包中包含系统所有数据库实体类。Control 包中包含所有系统资源,业务处理类和数据库连接类。

(2) 功能模块划分:图书管理员登录、读者管理、图书管理、图书类别设置、读者类别设置、管理员设置、借阅操作、罚款操作和信息统计。

(3) 数据库设计:各个数据库表的设计,以及表间的关联。

3. 详细设计小结

详细设计的主要任务是逐个对各个层次中的每个功能模块进行功能细节设计和系统调用函数功能设计,可以采用 IPO 图描述程序功能,完成程序的性能要求。采用 UML 类图,描述项目中用到的类的属性、方法、约束等详细内容。参照这些完成软件界面设计、模块数据表设计、类代码编写等工作,实现程序代码化。

通过对功能的实现,进一步提高学生对 Java 编程的熟练程度,加强学生对开发规范的认识,提高学生的分析问题和解决问题的能力,锻炼学生软件开发思维方式,增加学生的学习兴趣,引导学生就业前景的思考。

4. 项目测试

软件测试的主要任务是理解产品的功能要求,并对其进行测试,检查软件有没有错误,决定软件是否具有稳定性,写出软件测试相应的测试规范和测试用例。测试工作经过单元测试、集成测试、系统测试、验收测试和回归测试。

测试人员应完成的工作有6个。

(1) 编写软件测试计划,搭建测试环境。

(2) 编写软件测试用例,执行测试用例。

(3) 提交测试缺陷,跟踪缺陷以及编写测试报告。

(4) 准确地定位并跟踪问题,推动问题及时合理地解决。

(5) 制定性能测试方案、性能测试。

(6) 参与系统需求分析、设计、变更。

软件开发完成后通过对该系统进行全面仔细的检查测试,设计各种测试用例,有效地检查出系统存在的问题,进一步改进系统,提高软件质量,同时进一步提高学生的创新性和综合分析能力,具备判断准确、追求完美、执着认真、善于合作的品质,丰富编程经验,提高检查故障的能力。

5. 项目后记

至此,项目已经完成。通过本项目可以加深对分层概念的理解和学习,熟悉项目的总体结构和编写流程。在理解业务流程的基础上加深对控件的应用,熟练对数据库的相关操作,学习了ireport报表工具的使用,项目重点是培养学生进行需求分析设计能力、业务接受能力、编码能力的锻炼,希望此项目能为读者的Java学习有所帮助。